**NONRESIDENT TRAINING COURSE**

SEPTEMBER 1998

Navy Electricity and Electronics Training Series

Module 5—Introduction to Generators and Motors

NAVEDTRA 14177

DISTRIBUTION STATEMENT A: Approved for public release; distribution is unlimited.

Although the words "he," "him," and "his" are used sparingly in this course to enhance communication, they are not intended to be gender driven or to affront or discriminate against anyone.

DISTRIBUTION STATEMENT A: Approved for public release; distribution is unlimited.

PREFACE

By enrolling in this self-study course, you have demonstrated a desire to improve yourself and the Navy. Remember, however, this self-study course is only one part of the total Navy training program. Practical experience, schools, selected reading, and your desire to succeed are also necessary to successfully round out a fully meaningful training program.

COURSE OVERVIEW: To introduce the student to the subject of Generators and Motors who needs such a background in accomplishing daily work and/or in preparing for further study.

THE COURSE: This self-study course is organized into subject matter areas, each containing learning objectives to help you determine what you should learn along with text and illustrations to help you understand the information. The subject matter reflects day-to-day requirements and experiences of personnel in the rating or skill area. It also reflects guidance provided by Enlisted Community Managers (ECMs) and other senior personnel, technical references, instructions, etc., and either the occupational or naval standards, which are listed in the Manual of Navy Enlisted Manpower Personnel Classifications and Occupational Standards, NAVPERS 18068.

THE QUESTIONS: The questions that appear in this course are designed to help you understand the material in the text.

VALUE: In completing this course, you will improve your military and professional knowledge. Importantly, it can also help you study for the Navy-wide advancement in rate examination. If you are studying and discover a reference in the text to another publication for further information, look it up.

1998 Edition Prepared by FTCS(SS) Steven F. Reith

Published by NAVAL EDUCATION AND TRAINING PROFESSIONAL DEVELOPMENT AND TECHNOLOGY CENTER

NAVSUP Logistics Tracking Number 0504-LP-026-8300

Sailor's Creed

"I am a United States Sailor.

I will support and defend the Constitution of the United States of America and I will obey the orders of those appointed over me.

I represent the fighting spirit of the Navy and those who have gone before me to defend freedom and democracy around the world.

I proudly serve my country's Navy combat team with honor, courage and commitment.

I am committed to excellence and the fair treatment of all."

TABLE OF CONTENTS

CHAPTER PAGE

1. Direct Current Generators 1-1
2. Direct Current Motors 2-1
3. Alternating Current Generators 3-1

4. Alternating Current Motors 4-1

APPENDIX

I. Glossary AI-1

INDEX INDEX-1

NAVY ELECTRICITY AND ELECTRONICS TRAINING SERIES

The Navy Electricity and Electronics Training Series (NEETS) was developed for use by personnel in many electrical- and electronic-related Navy ratings. Written by, and with the advice of, senior technicians in these ratings, this series provides beginners with fundamental electrical and electronic concepts through self-study. The presentation of this series is not oriented to any specific rating structure, but is divided into modules containing related information organized into traditional paths of instruction.

The series is designed to give small amounts of information that can be easily digested before advancing further into the more complex material. For a student just becoming acquainted with electricity or electronics, it is highly recommended that the modules be studied in their suggested sequence. While there is a listing of NEETS by module title, the following brief descriptions give a quick overview of how the individual modules flow together.

Module 1, Introduction to Matter, Energy, and Direct Current, introduces the course with a short history of electricity and electronics and proceeds into the characteristics of matter, energy, and direct current (dc). It also describes some of the general safety precautions and first-aid procedures that should be common knowledge for a person working in the field of electricity. Related safety hints are located throughout the rest of the series, as well.

Module 2, Introduction to Alternating Current and Transformers, is an introduction to alternating current (ac) and transformers, including basic ac theory and fundamentals of electromagnetism, inductance, capacitance, impedance, and transformers.

Module 3, Introduction to Circuit Protection, Control, and Measurement, encompasses circuit breakers, fuses, and current limiters used in circuit protection, as well as the theory and use of meters as electrical measuring devices.

Module 4, Introduction to Electrical Conductors, Wiring Techniques, and Schematic Reading, presents conductor usage, insulation used as wire covering, splicing, termination of wiring, soldering, and reading electrical wiring diagrams.

Module 5, Introduction to Generators and Motors, is an introduction to generators and motors, and covers the uses of ac and dc generators and motors in the conversion of electrical and mechanical energies.

Module 6, Introduction to Electronic Emission, Tubes, and Power Supplies, ties the first five modules together in an introduction to vacuum tubes and vacuum-tube power supplies.

Module 7, Introduction to Solid-State Devices and Power Supplies, is similar to module 6, but it is in reference to solid-state devices.

Module 8, Introduction to Amplifiers, covers amplifiers.

Module 9, Introduction to Wave-Generation and Wave-Shaping Circuits, discusses wave generation and wave-shaping circuits.

Module 10, Introduction to Wave Propagation, Transmission Lines, and

Antennas, presents the characteristics of wave propagation, transmission lines, and antennas.

Module 11, Microwave Principles, explains microwave oscillators, amplifiers, and waveguides. Module 12, Modulation Principles, discusses the principles of modulation.

Module 13, Introduction to Number Systems and Logic Circuits, presents the fundamental concepts of number systems, Boolean algebra, and logic circuits, all of which pertain to digital computers.

Module 14, Introduction to Microelectronics, covers microelectronics technology and miniature and microminiature circuit repair.

Module 15, Principles of Synchros, Servos, and Gyros, provides the basic principles, operations, functions, and applications of synchro, servo, and gyro mechanisms.

Module 16, Introduction to Test Equipment, is an introduction to some of the more commonly used test equipments and their applications.

Module 17, Radio-Frequency Communications Principles, presents the fundamentals of a radio-frequency communications system.

Module 18, Radar Principles, covers the fundamentals of a radar system.

Module 19, The Technician's Handbook, is a handy reference of commonly used general information, such as electrical and electronic formulas, color coding, and naval supply system data.

Module 20, Master Glossary, is the glossary of terms for the series.

Module 21, Test Methods and Practices, describes basic test methods and practices.

Module 22, Introduction to Digital Computers, is an introduction to digital computers.

Module 23, Magnetic Recording, is an introduction to the use and maintenance of magnetic recorders and the concepts of recording on magnetic tape and disks.

Module 24, Introduction to Fiber Optics, is an introduction to fiber optics.

Embedded questions are inserted throughout each module, except for modules 19 and 20, which are reference books. If you have any difficulty in answering any of the questions, restudy the applicable section.

Although an attempt has been made to use simple language, various technical words and phrases have necessarily been included. Specific terms are defined in Module 20, Master Glossary.

Considerable emphasis has been placed on illustrations to provide a maximum amount of information. In some instances, a knowledge of basic algebra may be required.

Assignments are provided for each module, with the exceptions of Module 19, The Technician's Handbook; and Module 20, Master Glossary. Course descriptions and ordering information are in NAVEDTRA 12061, Catalog of Nonresident Training Courses.

Throughout the text of this course and while using technical manuals associated with the equipment you will be working on, you will find the below notations at the end of some paragraphs. The notations are used to emphasize that safety hazards exist and care must be taken or observed.

WARNING

AN OPERATING PROCEDURE, PRACTICE, OR CONDITION, ETC., WHICH MAY RESULT IN INJURY OR DEATH IF NOT CAREFULLY OBSERVED OR FOLLOWED.

CAUTION

AN OPERATING PROCEDURE, PRACTICE, OR CONDITION, ETC., WHICH MAY RESULT IN DAMAGE TO EQUIPMENT IF NOT CAREFULLY OBSERVED OR FOLLOWED.

NOTE

An operating procedure, practice, or condition, etc., which is essential to emphasize.

INSTRUCTIONS FOR TAKING THE COURSE

ASSIGNMENTS

The text pages that you are to study are listed at the beginning of each assignment. Study these pages carefully before attempting to answer the questions. Pay close attention to tables and illustrations and read the learning objectives. The learning objectives state what you should be able to do after studying the material. Answering the questions correctly helps you accomplish the objectives.

SELECTING YOUR ANSWERS

Read each question carefully, then select the BEST answer. You may refer freely to the text. The answers must be the result of your own work and decisions. You are prohibited from referring to or copying the answers of others and from giving answers to anyone else taking the course.

SUBMITTING YOUR ASSIGNMENTS

To have your assignments graded, you must be enrolled in the course with the Nonresident Training Course Administration Branch at the Naval Education and Training Professional Development and Technology Center (NETPDTC). Following enrollment, there are two ways of having your assignments graded: (1) use the Internet to submit your assignments as you complete them, or (2) send all the assignments at one time by mail to NETPDTC.

Grading on the Internet: Advantages to Internet grading are:

• you may submit your answers as soon as you complete an assignment, and

• you get your results faster; usually by the next working day (approximately 24 hours).

In addition to receiving grade results for each assignment, you will receive course completion confirmation once you have completed all the

assignments. To submit your assignment answers via the Internet, go to:

http ://courses.cnet.na vy.mil

Grading by Mail: When you submit answer sheets by mail, send all of your assignments at one time. Do NOT submit individual answer sheets for grading. Mail all of your assignments in an envelope, which you either provide yourself or obtain from your nearest Educational Services Officer (ESO). Submit answer sheets to:

COMMANDING OFFICER NETPDTC N331 6490 SAUFLEY FIELD ROAD PENSACOLA FL 32559-5000

Answer Sheets: All courses include one "scannable" answer sheet for each assignment. These answer sheets are preprinted with your SSN, name, assignment number, and course number. Explanations for completing the answer sheets are on the answer sheet.

Do not use answer sheet reproductions: Use only the original answer sheets that we provide—reproductions will not work with our scanning equipment and cannot be processed.

Follow the instructions for marking your answers on the answer sheet. Be sure that blocks 1, 2, and 3 are filled in correctly. This information is necessary for your course to be properly processed and for you to receive credit for your work.

COMPLETION TIME

Courses must be completed within 12 months from the date of enrollment. This includes time required to resubmit failed assignments.

PASS/FAIL ASSIGNMENT PROCEDURES

If your overall course score is 3.2 or higher, you will pass the course and will not be required to resubmit assignments. Once your assignments have been graded you will receive course completion confirmation.

If you receive less than a 3.2 on any assignment and your overall course score is below 3.2, you will be given the opportunity to resubmit failed assignments. You may resubmit failed assignments only once. Internet students will receive notification when they have failed an assignment—they may then resubmit failed assignments on the web site. Internet students may view and print results for failed assignments from the web site. Students who submit by mail will receive a failing result letter and a new answer sheet for resubmission of each failed assignment.

COMPLETION CONFIRMATION

After successfully completing this course, you will receive a letter of completion.

ERRATA

Errata are used to correct minor errors or delete obsolete information in a course. Errata may also be used to provide instructions to the student. If a course has an errata, it will be included as the first page(s) after the front cover. Errata for all courses can be accessed and viewed/downloaded at:

http://www.advancement.cnet.navy.mil

STUDENT FEEDBACK QUESTIONS

We value your suggestions, questions, and criticisms on our courses. If you would like to communicate with us regarding this course, we encourage you, if possible, to use e-mail. If you write or fax, please use a copy of the Student Comment form that follows this page.

For subject matter questions:

E-mail: n315.products@cnet.navy.mil Phone: Comm: (850) 452-1001, ext. 1728 DSN: 922-1001, ext. 1728 FAX: (850)452-1370 (Do not fax answer sheets.) Address: COMMANDING OFFICER NETPDTC N315 6490 SAUFLEY FIELD ROAD PENSACOLA FL 32509-5237

For enrollment, shipping, grading, or completion letter questions

E-mail: fleetservices @ cnet. navy. mil

Phone: Toll Free: 877-264-8583

Comm: (850)452-1511/1181/1859 DSN: 922-1511/1181/1859 FAX: (850)452-1370 (Do not fax answer sheets.)

Address: COMMANDING OFFICER NETPDTC N331 6490 SAUFLEY FIELD ROAD PENSACOLA FL 32559-5000

NAVAL RESERVE RETIREMENT CREDIT

If you are a member of the Naval Reserve, you will receive retirement points if you are authorized to receive them under current directives governing retirement of Naval Reserve personnel. For Naval Reserve retirement, this course is evaluated at 2 points. (Refer to Administrative Procedures for Naval Reservists on Inactive Duty, BUPERSINST 1001.39, for more information about retirement points.)

Student Comments

NEETS Module 5 Course Title: Introduction to Generators and Motors
NAVEDTRA: 14177 Date:
We need some information about you :
Rate/Rank and Name: SSN: Command/Unit
Street Address: City: State/FPO: Zip
Your comments, suggestions, etc.:

Privacy Act Statement: Under authority of Title 5, USC 301, information regarding your military status is requested in processing your comments and in preparing a reply. This information will not be divulged without written authorization to anyone other than those within POD for official use in determining performance.

NETPDTC 1550/41 (Rev 4-00)

CHAPTER 1

DIRECT CURRENT GENERATORS

LEARNING OBJECTIVES

Upon completion of the chapter you will be able to:

1. State the principle by which generators convert mechanical energy to electrical energy.
2. State the rule to be applied when you determine the direction of induced emf in a coil.
3. State the purpose of slip rings.
4. State the reason why no emf is induced in a rotating coil as it passes through a neutral plane.
5. State what component causes a generator to produce direct current rather than alternating current.
6. Identify the point at which the brush contact should change from one commutator segment to the next.
7. State how field strength can be varied in a dc generator.
8. Describe the cause of sparking between brushes and commutator.
9. State what is meant by "armature reaction."
10. State the purpose of interpoles.
11. Explain the effect of motor reaction in a dc generator.
12. Explain the causes of armature losses.
13. List the types of armatures used in dc generators.
14. State the three classifications of dc generators.
15. State the term that applies to voltage variation from no-load to full-load conditions and how it is expressed as a percentage.
16. State the term that describes the use of two or more generators to supply a common load.
17. State the purpose of a dc generator that has been modified to function as an

amplidyne.

INTRODUCTION

A generator is a machine that converts mechanical energy into electrical energy by using the principle of magnetic induction. This principle is explained as follows:

Whenever a conductor is moved within a magnetic field in such a way that the conductor cuts across magnetic lines of flux, voltage is generated in the conductor.

The AMOUNT of voltage generated depends on (1) the strength of the magnetic field, (2) the angle at which the conductor cuts the magnetic field, (3) the speed at which the conductor is moved, and (4) the length of the conductor within the magnetic field.

The POLARITY of the voltage depends on the direction of the magnetic lines of flux and the direction of movement of the conductor. To determine the direction of current in a given situation, the LEFT-HAND RULE FOR GENERATORS is used. This rule is explained in the following manner.

Extend the thumb, forefinger, and middle finger of your left hand at right angles to one another, as shown in figure 1-1. Point your thumb in the direction the conductor is being moved. Point your forefinger in the direction of magnetic flux (from north to south). Your middle finger will then point in the direction of current flow in an external circuit to which the voltage is applied.

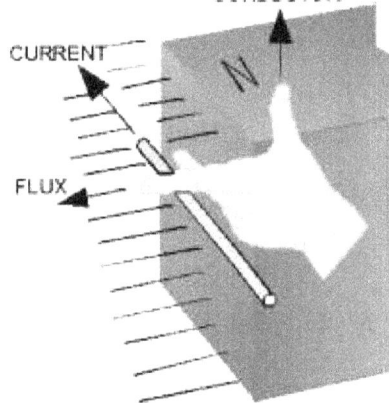

Figure 1-1.—Left-hand rule for generators.

THE ELEMENTARY GENERATOR

The simplest elementary generator that can be built is an ac generator. Basic generating principles are most easily explained through the use of the elementary ac generator. For this reason, the ac generator will be discussed first. The dc generator will be discussed later.

An elementary generator (fig. 1-2) consists of a wire loop placed so that it can be rotated in a stationary magnetic field. This will produce an induced emf in the loop. Sliding contacts (brushes) connect the loop to an external circuit load in order to pick up or use the induced emf.

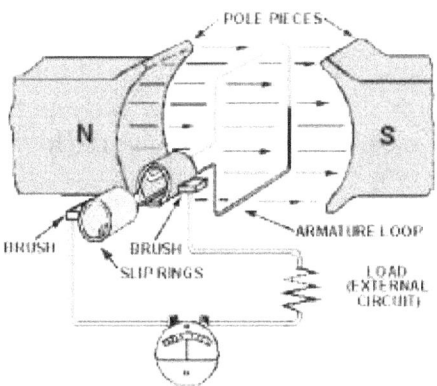

Figure 1-2.—The elementary generator.

The pole pieces (marked N and S) provide the magnetic field. The pole pieces are shaped and positioned as shown to concentrate the magnetic field as close as possible to the wire loop. The loop of wire that rotates through the field is called the ARMATURE. The ends of the armature loop are connected to rings called SLIP RINGS. They rotate with the armature. The brushes, usually made of carbon, with wires attached to them, ride against the rings. The generated voltage appears across these brushes.

The elementary generator produces a voltage in the following manner (fig. 1-3). The armature loop is rotated in a clockwise direction. The initial or starting point is shown at position A. (This will be considered the zero-degree position.) At 0° the armature loop is perpendicular to the magnetic field. The black and white conductors of the loop are moving parallel to the field. The instant the conductors are moving parallel to the magnetic field, they do not cut any lines of flux. Therefore, no emf is induced in the conductors, and the meter at position A indicates zero. This position is called the NEUTRAL PLANE. As the armature loop rotates from position A (0°) to position B (90°), the conductors cut through more and more lines of flux, at a continually increasing angle. At 90° they are cutting through a maximum number of lines of flux and at maximum angle. The result is that between 0° and 90°, the induced emf in the conductors builds up from zero to a maximum value. Observe that from 0° to 90°, the black conductor cuts DOWN through the field. At the same time the white conductor cuts UP through the field. The induced emfs in the conductors are series-adding. This means the resultant voltage across the brushes (the terminal voltage) is the sum of the two induced voltages. The meter at position B reads maximum value. As the armature loop continues rotating from 90° (position B) to 180° (position C), the conductors which were cutting through a maximum number of lines of flux at position B now cut through fewer lines. They are again moving parallel to the magnetic field at position C. They no longer cut through any lines of flux. As the armature rotates from 90° to 180°, the induced voltage will decrease to zero in the same manner that it increased during the rotation from 0° to 90°. The meter again reads zero. From 0° to 180° the conductors of the armature loop have been moving in the same direction through the magnetic field. Therefore, the polarity of the induced voltage has remained the same. This is shown by points A through C on the graph. As the loop rotates beyond 180° (position C), through 270° (position D), and back to the initial or starting point (position A), the direction of the cutting action of the conductors through the magnetic field reverses. Now the black conductor cuts UP through the field while the white conductor cuts DOWN through the field. As a result, the polarity of the induced voltage reverses. Following the sequence shown by graph points

C, D, and back to A, the voltage will be in the direction opposite to that shown from points A, B, and C. The terminal voltage will be the same as it was from A to C except that the polarity is reversed (as shown by the meter deflection at position D). The voltage output waveform for the complete revolution of the loop is shown on the graph in figure 1-3.

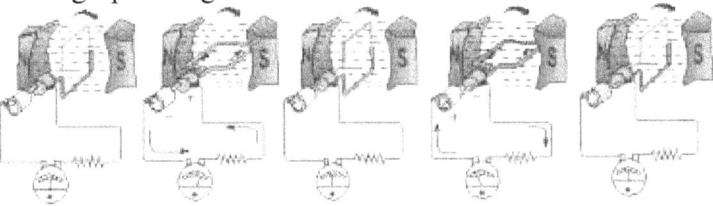

A B C D A
(0 &) (90<) (1804 (270 <J
POSITION POSITION POSITION POSITION POSITION
A B C D A
GENERATOR TERMINAL VOLTAGE

Figure 1-3.—Output voltage of an elementary generator during one revolution.

Q1. Generators convert mechanical motion to electrical energy using what principle? Q2. What rule should you use to determine the direction of induced emf in a coil? Q3. What is the purpose of the slip rings?

Q4. Why is no emf induced in a rotating coil when it passes through the neutral plane?

THE ELEMENTARY DC GENERATOR

A single-loop generator with each terminal connected to a segment of a two-segment metal ring is shown in figure 1-4. The two segments of the split metal ring are insulated from each other. This forms a simple COMMUTATOR. The commutator in a dc generator replaces the slip rings of the ac generator. This is the main difference in their construction. The commutator mechanically reverses the armature loop connections to the external circuit. This occurs at the same instant that the polarity of the voltage in the armature loop reverses. Through this process the commutator changes the generated ac voltage to a pulsating dc voltage as shown in the graph of figure 1-4. This action is known as commutation. Commutation is described in detail later in this chapter.

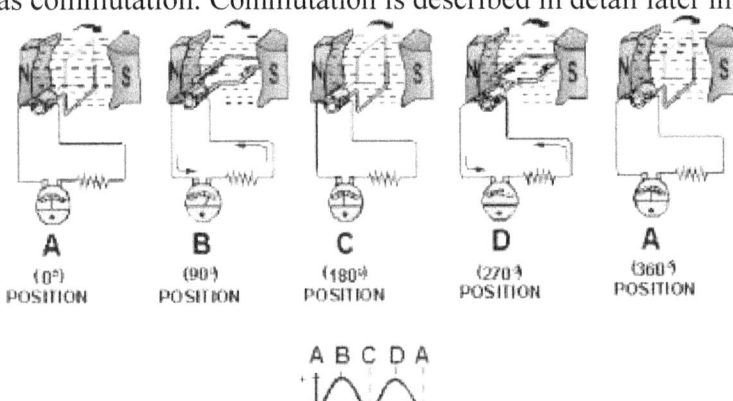

Figure 1-4.—Effects of commutation.

For the remainder of this discussion, refer to figure 1-4, parts A through D. This will help you in following the step-by-step description of the operation of a dc generator. When the armature loop rotates clockwise from position A to position B, a voltage is induced in the armature loop which causes a current in a direction that deflects the meter to the right. Current flows through loop, out of the negative brush, through the meter and the load, and back through the positive brush to the loop. Voltage reaches its maximum value at point B on the graph for reasons explained earlier. The generated voltage and the current fall to zero at position C. At this instant each brush makes contact with both segments of the commutator. As the armature loop rotates to position D, a voltage is again induced in the loop. In this case, however, the voltage is of opposite polarity.

The voltages induced in the two sides of the coil at position D are in the reverse direction to that of the voltages shown at position B. Note that the current is flowing from the black side to the white side in position B and from the white side to the black side in position D. However, because the segments of the commutator have rotated with the loop and are contacted by opposite brushes, the direction of current flow through the brushes and the meter remains the same as at position B. The voltage developed across the brushes is pulsating and unidirectional (in one direction only). It varies twice during each revolution between zero and maximum. This variation is called RIPPLE.

A pulsating voltage, such as that produced in the preceding description, is unsuitable for most applications. Therefore, in practical generators more armature loops (coils) and more commutator segments are used to produce an output voltage waveform with less ripple.

Q5. What component causes a generator to produce dc voltage rather than ac voltage at its output terminals?

Q6. At what point should brush contact change from one commutator segment to the next?

Q7. An elementary, single coil, dc generator will have an output voltage with how many pulsations per revolution?

EFFECTS OF ADDING ADDITIONAL COILS AND POLES

The effects of additional coils may be illustrated by the addition of a second coil to the armature. The commutator must now be divided into four parts since there are four coil ends (see fig. 1-5). The coil is rotated in a clockwise direction from the position shown. The voltage induced in the white coil, DECREASES FOR THE NEXT 90° of rotation (from maximum to zero). The voltage induced in the black coil INCREASES from zero to maximum at the same time. Since there are four segments in the commutator, a new segment passes each brush every 90° instead of every 180°. This allows the brush to switch from the white coil to the black coil at the instant the voltages in the two coils are equal. The brush remains in contact with the black coil as its induced voltage increases to maximum, level B in the graph. It then decreases to level A, 90° later. At this point, the brush will contact the white coil again.

POLE PIECES

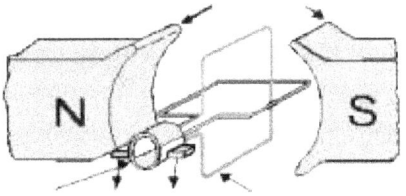

COMMUTATOR TWO-COIL ARMATURE
GENERATOR TERMINAL VOLTAGE

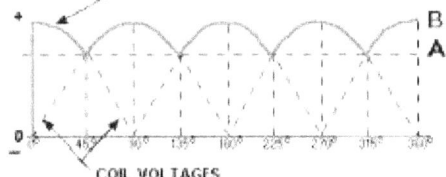

Figure 1-5.—Effects of additional coils.

The graph in figure 1-5 shows the ripple effect of the voltage when two armature coils are used. Since there are now four commutator segments in the commutator and only two brushes, the voltage cannot fall any lower than at point A. Therefore, the ripple is limited to the rise and fall between points A and B on the graph. By adding more armature coils, the ripple effect can be further reduced. Decreasing ripple in this way increases the effective voltage of the output.

NOTE: Effective voltage is the equivalent level of dc voltage, which will cause the same average current through a given resistance. By using additional armature coils, the voltage across the brushes is not allowed to fall to as low a level between peaks. Compare the graphs in figure 1-4 and 1-5. Notice that the ripple has been reduced. Practical generators use many armature coils. They also use more than one pair of magnetic poles. The additional magnetic poles have the same effect on ripple as did the additional armature coils. In addition, the increased number of poles provides a stronger magnetic field (greater number of flux lines). This, in turn, allows an increase in output voltage because the coils cut more lines of flux per revolution.

Q8. How many commutator segments are required in a two-coil generator?

ELECTROMAGNETIC POLES

Nearly all practical generators use electromagnetic poles instead of the permanent magnets used in our elementary generator. The electromagnetic field poles consist of coils of insulated copper wire wound on soft iron cores, as shown in figure 1-6. The main advantages of using electromagnetic poles are (1) increased field strength and (2) a means of controlling the strength of the fields. By varying the input voltage, the field strength is varied. By varying the field strength, the output voltage of the generator can be controlled.

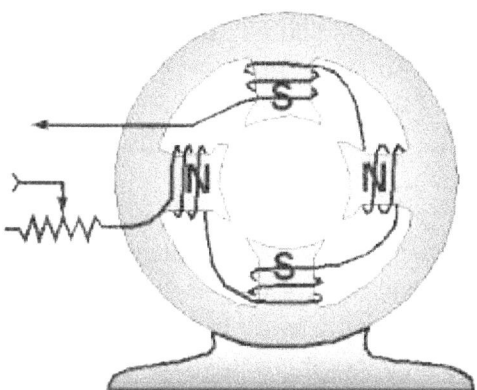

Figure 1-6.—Four-pole generator (without armature).

Q9. How can field strength be varied in a practical dc generator?

COMMUTATION

Commutation is the process by which a dc voltage output is taken from an armature that has an ac voltage induced in it. You should remember from our discussion of the elementary dc generator that the commutator mechanically reverses the armature loop connections to the external circuit. This occurs at the same instant that the voltage polarity in the armature loop reverses. A dc voltage is applied to the load because the output connections are reversed as each commutator segment passes under a brush. The segments are insulated from each other.

In figure 1-7, commutation occurs simultaneously in the two coils that are briefly short-circuited by the brushes. Coil B is short-circuited by the negative brush. Coil Y, the opposite coil, is short-circuited by the positive brush. The brushes are positioned on the commutator so that each coil is short-circuited as it moves through its own electrical neutral plane. As you have seen previously, there is no voltage generated in the coil at that time. Therefore, no sparking can occur between the commutator and the brush. Sparking between the brushes and the commutator is an indication of improper commutation. Improper brush placement is the main cause of improper commutation.

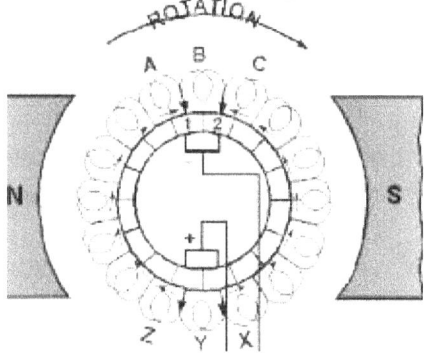

LOAD

Figure 1-7.—Commutation of a dc generator.

Q10. What causes sparking between the brushes and the commutator?

ARMATURE REACTION

From previous study, you know that all current-carrying conductors produce magnetic fields. The magnetic field produced by current in the armature of a dc generator affects the flux pattern and distorts the main field. This distortion causes a shift in the neutral plane, which affects commutation. This change in the neutral plane and the reaction of the magnetic field is called ARMATURE REACTION.

You know that for proper commutation, the coil short-circuited by the brushes must be in the neutral plane. Consider the operation of a simple two-pole dc generator, shown in figure 1-8. View A of the figure shows the field poles and the main magnetic field. The armature is shown in a simplified view in views B and C with the cross section of its coil represented as little circles. The symbols within the circles represent arrows. The dot represents the point of the arrow coming toward you, and the cross represents the tail, or feathered end, going away from you. When the armature rotates clockwise, the sides of the coil to the left will have current flowing toward you, as indicated by the dot. The side of the coil to the right will have current flowing away from you, as indicated by the cross. The field generated around each side of the coil is shown in view B of figure 1-8. This field increases in strength for each wire in the armature coil, and sets up a magnetic field almost perpendicular to the main field.

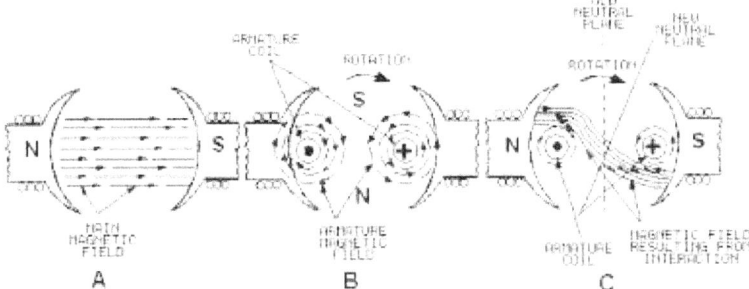

Figure 1-8.—Armature reaction.

Now you have two fields — the main field, view A, and the field around the armature coil, view B. View C of figure 1-8 shows how the armature field distorts the main field and how the neutral plane is shifted in the direction of rotation. If the brushes remain in the old neutral plane, they will be short-circuiting coils that have voltage induced in them. Consequently, there will be arcing between the brushes and commutator.

To prevent arcing, the brushes must be shifted to the new neutral plane.

Q1 1. What is armature reaction?

COMPENSATING WINDINGS AND INTERPOLES

Shifting the brushes to the advanced position (the new neutral plane) does not completely solve the problems of armature reaction. The effect of armature reaction varies with the load current. Therefore, each time the load current varies, the neutral plane shifts. This means the brush position must be changed each time the load current varies.

In small generators, the effects of armature reaction are reduced by actually mechanically shifting the position of the brushes. The practice of shifting the brush position for each current variation is not practiced except in small generators. In larger generators, other means are taken to eliminate armature reaction. COMPENSATING WINDINGS or INTERPOLES are used for this purpose (fig. 1-9). The compensating windings consist of a series of coils embedded in slots in the pole faces. These coils are connected in series with the armature. The series-connected compensating windings produce a magnetic field, which varies directly with armature current. Because the compensating windings are wound to produce a field that opposes the magnetic field of the armature, they tend to cancel the effects of the armature magnetic field. The neutral plane will remain stationary and in its original position for all values of armature current.

Because of this, once the brushes have been set correctly, they do not have to be moved again.

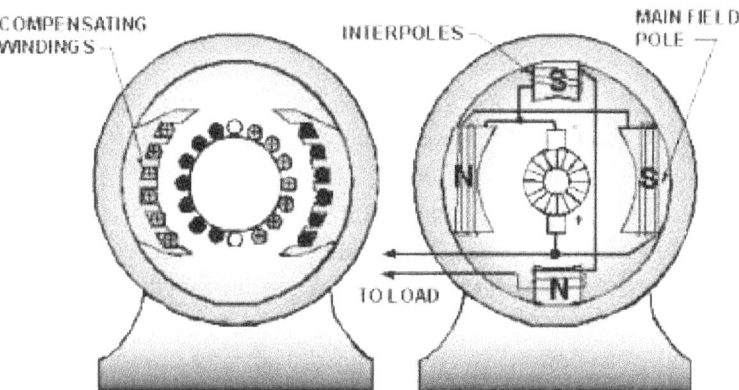

Figure 1-9.—Compensating windings and interpoles.

Another way to reduce the effects of armature reaction is to place small auxiliary poles called "interpoles" between the main field poles. The interpoles have a few turns of large wire and are connected in series with the armature. Interpoles are wound and placed so that each interpole has the same magnetic polarity as the main pole ahead of it, in the direction of rotation. The field generated by the interpoles produces the same effect as the compensating winding. This field, in effect, cancels the armature reaction for all values of load current by causing a shift in the neutral plane opposite to the shift caused by armature reaction. The amount of shift caused by the interpoles will equal the shift caused by armature reaction since both shifts are a result of armature current.

Q12. What is the purpose of interpoles?

MOTOR REACTION IN A GENERATOR

When a generator delivers current to a load, the armature current creates a magnetic force that opposes the rotation of the armature. This is called MOTOR REACTION. A single armature conductor is represented in figure 1-10, view A. When the conductor is stationary, no voltage is generated and no current flows. Therefore, no force acts on the conductor. When the conductor is moved downward (fig. 1-10, view B) and the circuit is completed through an external load, current flows through the conductor in the direction indicated. This sets up lines of flux around the conductor in a clockwise direction.

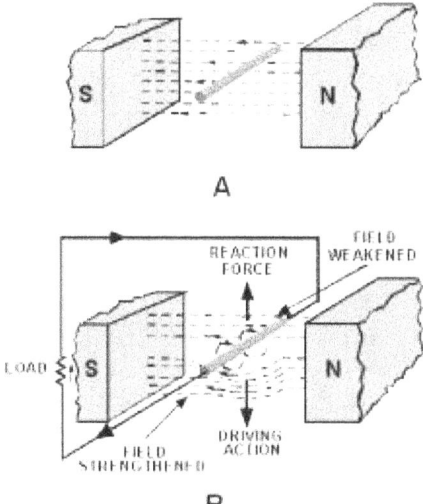

Figure 1-10.—Motor reaction in a generator.

The interaction between the conductor field and the main field of the generator weakens the field above the conductor and strengthens the field below the conductor. The main field consists of lines that now act like stretched rubber bands. Thus, an upward reaction force is produced that acts in opposition to the downward driving force applied to the armature conductor. If the current in the conductor increases, the reaction force increases. Therefore, more force must be applied to the conductor to keep it moving.

With no armature current, there is no magnetic (motor) reaction. Therefore, the force required to turn the armature is low. As the armature current increases, the reaction of each armature conductor against rotation increases. The actual force in a generator is multiplied by the number of conductors in the armature. The driving force required to maintain the generator armature speed must be increased to overcome the motor reaction. The force applied to turn the armature must overcome the motor reaction force in all dc generators. The device that provides the turning force applied to the armature is called the PRIME MOVER. The prime mover may be an electric motor, a gasoline engine, a steam turbine, or any other mechanical device that provides turning force.

Q13. What is the effect of motor reaction in a dc generator?

ARMATURE LOSSES

In dc generators, as in most electrical devices, certain forces act to decrease the efficiency. These forces, as they affect the armature, are considered as losses and may be defined as follows:

1. I^2R, or copper loss in the winding
2. Eddy current loss in the core
3. Hysteresis loss (a sort of magnetic friction)

Copper Losses

The power lost in the form of heat in the armature winding of a generator is known as COPPER LOSS. Heat is generated any time current flows in a conductor. Copper loss is an I^2R loss, which increases as current increases. The amount of heat generated is also proportional to the resistance of the conductor. The resistance of the conductor varies directly with its length and inversely with its cross-sectional area. Copper loss is minimized in armature windings by using large diameter wire.

Q14. What causes copper losses?

Eddy Current Losses

The core of a generator armature is made from soft iron, which is a conducting material with desirable magnetic characteristics. Any conductor will have currents induced in it when it is rotated in a magnetic field. These currents that are induced in the generator armature core are called EDDY CURRENTS. The power dissipated in the form of heat, as a result of the eddy currents, is considered a loss.

Eddy currents, just like any other electrical currents, are affected by the resistance of the material in which the currents flow. The resistance of any material is inversely proportional to its cross-sectional area. Figure 1-11, view A, shows the eddy currents induced in an armature core that is a solid piece of soft iron. Figure 1-11, view B, shows a soft iron core of the same size, but made up of several small pieces insulated from each other. This process is called lamination. The currents in each piece of the laminated core are considerably less than in the solid core because the resistance of the pieces is much higher. (Resistance is inversely proportional to cross-sectional area.) The currents in the individual pieces of the laminated core are so small that the sum of the individual currents is much less than the total of eddy currents in the solid iron core.

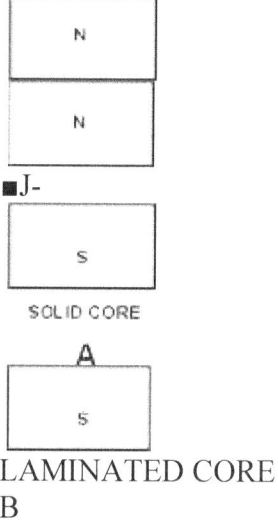

LAMINATED CORE
B

Figure 1-11.—Eddy currents in dc generator armature cores.

As you can see, eddy current losses are kept low when the core material is made up of many thin sheets of metal. Laminations in a small generator armature may be as thin as 1/64 inch. The laminations are insulated from each other by a thin coat of lacquer or, in some instances, simply by the oxidation of the surfaces. Oxidation is caused by contact with the air while the laminations are being annealed. The insulation value need not be high because the voltages induced are very small.

Most generators use armatures with laminated cores to reduce eddy current losses.

Q15. How can eddy current be reduced?

Hysteresis Losses

Hysteresis loss is a heat loss caused by the magnetic properties of the armature. When an armature core is in a magnetic field, the magnetic particles of the core tend to line up with the magnetic field. When the armature core is rotating, its magnetic field keeps changing direction. The continuous movement of the magnetic particles, as they try to align themselves with the magnetic field, produces molecular friction. This, in turn, produces heat. This heat is transmitted to the armature windings. The heat causes armature resistances to increase.

To compensate for hysteresis losses, heat-treated silicon steel laminations are used in most dc generator armatures. After the steel has been formed to the proper shape, the laminations are heated and allowed to cool. This annealing process reduces the hysteresis loss to a low value.

THE PRACTICAL DC GENERATOR

The actual construction and operation of a practical dc generator differs somewhat from our elementary generators. The differences are in the construction of the armature, the manner in which the armature is wound, and the method of developing the main field.

A generator that has only one or two armature loops has high ripple voltage. This results in too little current to be of any practical use. To increase the amount of current output, a number of loops of wire are used. These additional loops do away with most of the ripple. The loops of wire, called windings, are evenly spaced around the armature so that the distance between each winding is the same.

The commutator in a practical generator is also different. It has several segments instead of two or four, as in our elementary generators. The number of segments must equal the number of armature coils.

GRAMME-RING ARMATURE

The diagram of a GRAMME-RING armature is shown in figure 1-12, view A. Each coil is connected to two commutator segments as shown. One end of coil 1 goes to segment A, and the other end of coil 1 goes to segment B. One end of coil 2 goes to segment C, and the other end of coil 2 goes to segment B. The rest of the coils are connected in a like manner, in series, around the armature. To complete the series arrangement, coil 8 connects to segment A. Therefore, each coil is in series with every other coil.

NEUTRAL PLANE
I

[Figure showing Gramme-ring armature with CURRENT IN, DIRECTION OF ROTATION, ARMATURE WINDINGS, N and S poles, COMMUTATOR]

A. END VIEW B. COMPOSITE VIEW

Figure 1-12.—Gramme-ring armature.

Figure 1-12, view B shows a composite view of a Gramme-ring armature. It illustrates more graphically the physical relationship of the coils and commutator locations.

The windings of a Gramme-ring armature are placed on an iron ring. A disadvantage of this arrangement is that the windings located on the inner side of the iron ring cut few lines of flux. Therefore, they have little, if any, voltage induced in them. For this reason, the Gramme-ring armature is not widely used.

DRUM-TYPE ARMATURE

A drum-type armature is shown in figure 1-13. The armature windings are placed in slots cut in a drum-shaped iron core. Each winding completely surrounds the core so that the entire length of the conductor cuts the main magnetic field. Therefore, the total voltage induced in the armature is greater than in the Gramme-ring. You can see that the drum-type armature is much more efficient than the Gramme-ring. This accounts for the almost universal use of the drum-type armature in modem dc generators.

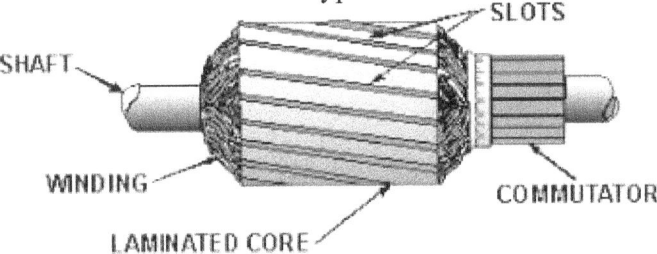

Figure 1-13.—Drum-type armature.

Drum-type armatures are wound with either of two types of windings — the LAP

WINDING or the WAVE WINDING.

The lap winding is illustrated in figure 1-14, view A This type of winding is used in dc generators designed for high-current applications. The windings are connected to provide several parallel paths for current in the armature. For this reason, lap-wound armatures used in dc generators require several pairs of poles and brushes.

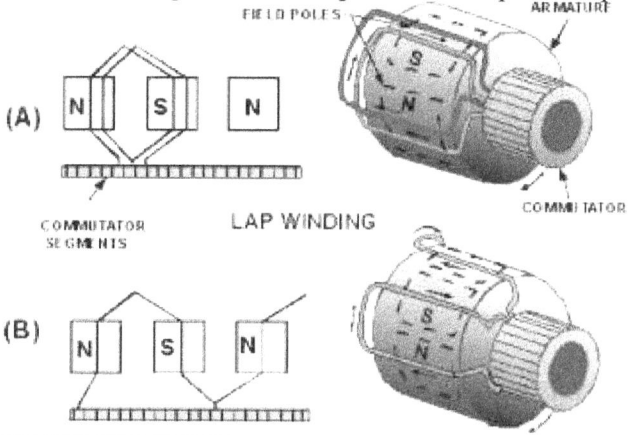

WAVE WINDING

Figure 1-14.—Types of windings used on drum-type armatures.

Figure 1-14, view B, shows a wave winding on a drum-type armature. This type of winding is used in dc generators employed in high-voltage applications. Notice that the two ends of each coil are connected to commutator segments separated by the distance between poles. This configuration allows the series addition of the voltages in all the windings between brushes. This type of winding only requires one pair of brushes. In practice, a practical generator may have several pairs to improve commutation.

Q16. Why are drum-type armatures preferred over the Gramme-ring armature in modern dc generators?

Q17. Lap windings are used in generators designed for what type of application?

FIELD EXCITATION

When a dc voltage is applied to the field windings of a dc generator, current flows through the windings and sets up a steady magnetic field. This is called FIELD EXCITATION.

This excitation voltage can be produced by the generator itself or it can be supplied by an outside source, such as a battery. A generator that supplies its own field excitation is called a SELF-EXCITED GENERATOR. Self-excitation is possible only if the field pole pieces have retained a slight amount of permanent magnetism, called RESIDUAL MAGNETISM. When the generator starts rotating, the weak residual magnetism causes a small voltage to be generated in the armature. This small voltage applied to

the field coils causes a small field current. Although small, this field current strengthens the magnetic field and allows the armature to generate a higher voltage. The higher voltage increases the field strength, and so on. This process continues until the output voltage reaches the rated output of the generator.

CLASSIFICATION OF GENERATORS

Self-excited generators are classed according to the type of field connection they use. There are three general types of field connections — SERIES-WOUND, SHUNT-WOUND (parallel), and COMPOUND-WOUND. Compound-wound generators are further classified as cumulative-compound and differential-compound. These last two classifications are not discussed in this chapter.

Series-Wound Generator

hi the series-wound generator, shown in figure 1-15, the field windings are connected in series with the armature. Current that flows in the armature flows through the external circuit and through the field windings. The external circuit connected to the generator is called the load circuit.

A series-wound generator uses very low resistance field coils, which consist of a few turns of large diameter wire.

The voltage output increases as the load circuit starts drawing more current. Under low-load current conditions, the current that flows in the load and through the generator is small. Since small current means that a small magnetic field is set up by the field poles, only a small voltage is induced in the armature. If the resistance of the load decreases, the load current increases. Under this condition, more current flows through the field. This increases the magnetic field and increases the output voltage. A series-wound dc generator has the characteristic that the output voltage varies with load current. This is undesirable in most applications. For this reason, this type of generator is rarely used in everyday practice.

The series-wound generator has provided an easy method to introduce you to the subject of self-excited generators.

SERIES FIELD
TO LOAD CIRCUIT

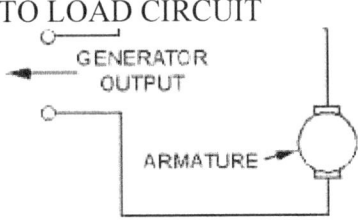

GENERATOR OUTPUT

ARMATURE

Figure 1-15.—Series-wound generator.

Shunt-Wound Generators

In a shunt-wound generator, like the one shown in figure 1-16, the field coils consist of many turns of small wire. They are connected in parallel with the load. In other words, they are connected across the output voltage of the armature.

GENERATOR OUTPUT
SHUNT FIELD
/
ARMATURE

6

Figure 1-16.—Shunt-wound generator.

Current in the field windings of a shunt-wound generator is independent of the load current (currents in parallel branches are independent of each other). Since field current, and therefore field strength, is not affected by load current, the output voltage remains more nearly constant than does the output voltage of the series-wound generator.

In actual use, the output voltage in a dc shunt-wound generator varies inversely as load current varies. The output voltage decreases as load current increases because the voltage drop across the armature resistance increases ($E = IR$).

In a series-wound generator, output voltage varies directly with load current. In the shunt-wound generator, output voltage varies inversely with load current. A combination of the two types can overcome the disadvantages of both. This combination of windings is called the compound-wound dc generator.

Compound-Wound Generators

Compound-wound generators have a series-field winding in addition to a shunt-field winding, as shown in figure 1-17. The shunt and series windings are wound on the same pole pieces.

GENERATOR OUTPUT
O—
SERIES FIELD
SHUNT FIELD
ARMATURE'

Figure 1-17.—Compound-wound generator.

In the compound-wound generator when load current increases, the armature voltage decreases just as in the shunt-wound generator. This causes the voltage applied to the shunt-field winding to decrease, which results in a decrease in the magnetic field. This same increase in load current, since it flows through the series winding, causes an increase in the magnetic field produced by that winding.

By proportioning the two fields so that the decrease in the shunt field is just compensated by the increase in the series field, the output voltage remains constant. This is shown in figure 1-18, which shows the voltage characteristics of the series-, shunt-, and compound-wound generators. As you can see, by proportioning the effects of the two fields (series and shunt), a compound-wound generator provides a constant output voltage under varying load conditions. Actual curves are seldom, if ever, as perfect as shown.

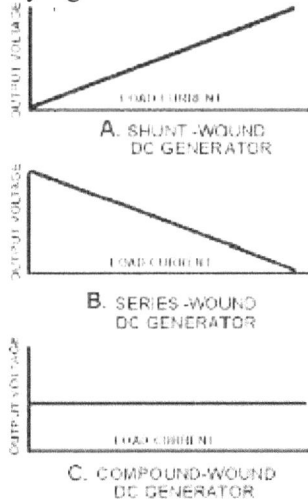

Figure 1-18.—Voltage output characteristics of the series-, shunt-, and compound-wound dc generators.

Q18. What are the three classifications of dc generators?

Q19. What is the main disadvantage of series generators?

GENERATOR CONSTRUCTION

Figure 1-19, views A through E, shows the component parts of dc generators. Figure 1-20 shows the entire generator with the component parts installed. The cutaway drawing helps you to see the physical relationship of the components to each other.

ARMATURE COILS

COMMUTATOR
TEETH
ARMAT

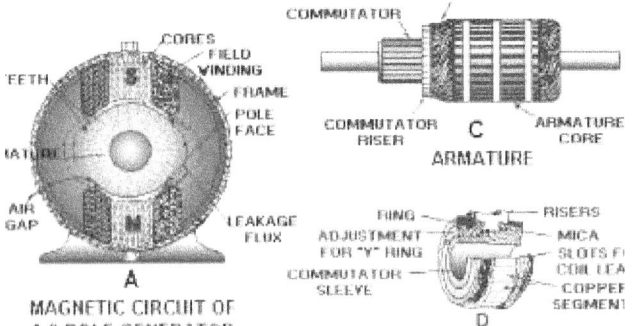

MAGNETIC CIRCUIT OF A 2-POLE GENERATOR
RISERS MICA
SLOTS FOR COIL LEADS COPPER SEGMENTS
COMMUTATOR CONSTRUCTION
CORE

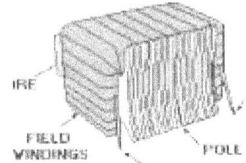

ADJUSTMENT FOR SPRING TENSION
LEAD
POLE FACE
BRUSH
LEAD
B
FIELD WINDINGS ON POLE PIECE

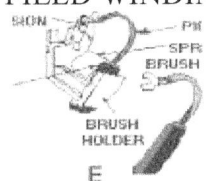

PIGTAIL
SPRING FOR BRUSH PRESSURE
TYPICAL PIGTAIL BRUSH AND HOLDER

Figure 1-19.—Components of a dc generator.

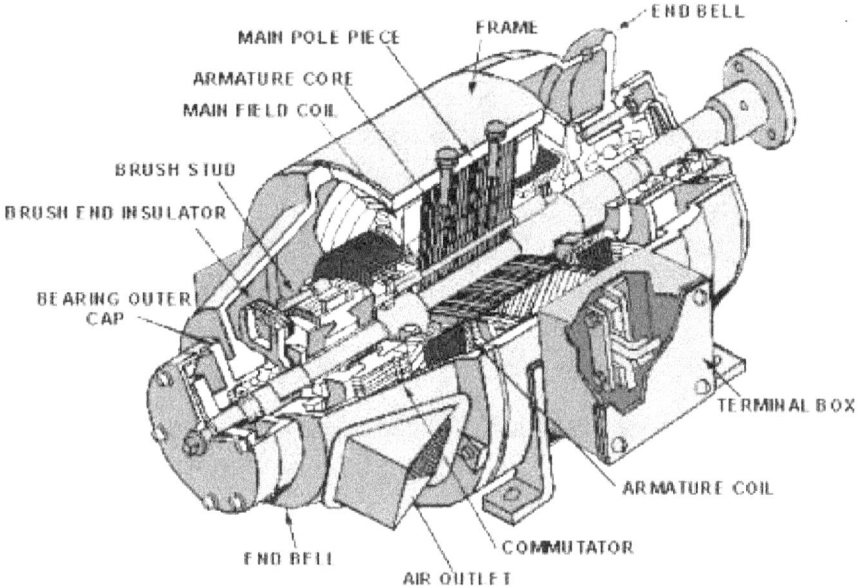

Figure 1-20.—Construction of a dc generator (cutaway drawing).

VOLTAGE REGULATION

The regulation of a generator refers to the VOLTAGE CHANGE that takes place when the load changes. It is usually expressed as the change in voltage from a no-load condition to a full-load condition, and is expressed as a percentage of full-load. It is expressed in the following formula:

Percent of reflation = — x 100
E fL

where EnL is the no-load terminal voltage and En. is the full-load terminal voltage of the generator. For example, to calculate the percent of regulation of a generator with a no-load voltage of 462 volts and a full-load voltage of 440 volts

Given:
• No-load voltage 462 V
• Full-load voltage 440 V Solution:

(E t - E*T)
Percent of regulation = — x 100
E fL

♦ r i (462V-440V)
Percent of regulation = x 100
440V

22V
Percent of regulation = x 100
440V

Percent of regulation =.05 x 100 Regulation = 5%

NOTE: The lower the percent of regulation, the better the generator. In the above example, the 5% regulation represented a 22-volt change from no load to full load. A 1 % change would represent a change of 4.4 volts, which, of course, would be better.

Q20. What term applies to the voltage variation from no-load to full-load conditions and is expressed as a percentage?

VOLTAGE CONTROL

Voltage control is either (1) manual or (2) automatic. In most cases the process involves changing the resistance of the field circuit. By changing the field circuit resistance, the field current is controlled. Controlling the field current permits control of the output voltage. The major difference between the various voltage control systems is merely the method by which the field circuit resistance and the current are controlled.

VOLTAGE REGULATION should not be confused with VOLTAGE CONTROL. As described previously, voltage regulation is an internal action occurring within the generator whenever the load changes. Voltage control is an imposed action, usually through an external adjustment, for the purpose of increasing or decreasing terminal voltage.

Manual Voltage Control

The hand-operated field rheostat, shown in figure 1-21, is a typical example of manual voltage control. The field rheostat is connected in series with the shunt field circuit. This provides the simplest method of controlling the terminal voltage of a dc generator.

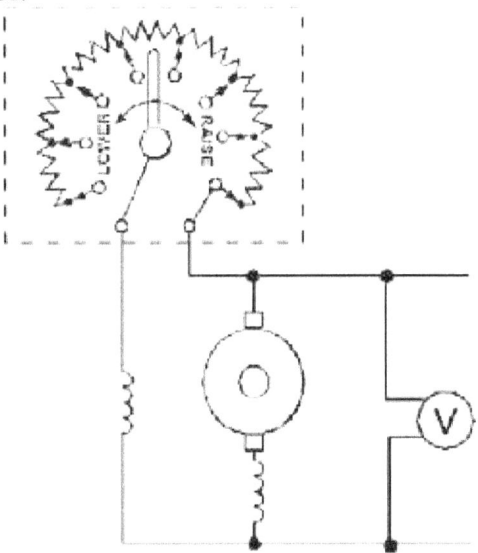

Figure 1-21.—Hand-operated field rheostat.

This type of field rheostat contains tapped resistors with leads to a multiterminal switch. The arm of the switch may be rotated to make contact with the various resistor taps. This varies the amount of resistance in the field circuit. Rotating the arm in the direction of the LOWER arrow (counterclockwise) increases the resistance and lowers the output voltage. Rotating the arm in the direction of the RAISE arrow (clockwise) decreases the resistance and increases the output voltage.

Most field rheostats for generators use resistors of alloy wire. They have a high specific resistance and a low temperature coefficient. These alloys include copper, nickel, manganese, and chromium. They are marked under trade names such as Nichrome, Advance, Manganin, and so forth. Some very large generators use cast-iron grids in place of rheostats, and motor-operated switching mechanisms to provide voltage control.

Automatic Voltage Control

Automatic voltage control may be used where load current variations exceed the built-in ability of the generator to regulate itself. An automatic voltage control device "senses" changes in output voltage and causes a change in field resistance to keep output voltage constant.

The actual circuitry involved in automatic voltage control will not be covered in this chapter. Whichever control method is used, the range over which voltage can be changed is a design characteristic of the generator. The voltage can be controlled only within the design limits.

PARALLEL OPERATION OF GENERATORS

When two or more generators are supplying a common load, they are said to be operating in parallel. The purpose of connecting generators in parallel is simply to provide more current than a single generator is capable of providing. The generators may be physically located quite a distance apart. However, they are connected to the common load through the power distribution system.

There are several reasons for operating generators in parallel. The number of generators used may be selected in accordance with the load demand. By operating each generator as nearly as possible to its rated capacity, maximum efficiency is achieved. A disabled or faulty generator may be taken off-line and replaced without interrupting normal operations.

Q21. What term applies to the use of two or more generators to supply a common load?

AMPLIDYNES

Amplidynes are special-purpose dc generators. They supply large dc currents, precisely controlled, to the large dc motors used to drive heavy physical loads, such as gun turrets and missile launchers.

The amplidyne is really a motor and a generator. It consists of a constant-speed ac motor (the prime mover) mechanically coupled to a dc generator, which is wired to function as a high-gain amplifier (an amplifier is a device in which a small input voltage can control a large current source). For instance, in a normal dc generator, a small dc voltage applied to the field windings is able to control the output of the generator. In a typical generator, a change in voltage from 0-volt dc to 3-volts dc applied to the field winding may cause the generator output to vary from 0-volt dc to 300-volts dc. If the 3 volts applied to the field winding is considered an input, and the 300 volts taken from the brushes is an output, there is a gain of 100. Gain is expressed as the ratio of output to input:

„ . output Gam = ——— input

In this case 300 V -s- 3 V = 100. This means that the 3 volts output is 100 times larger than the input.

The following paragraphs explain how gain is achieved in a typical dc generator and how the modifications making the generator an amplidyne increase the gain to as high as 10,000.

The schematic diagram in figure 1-22 shows a separately excited dc generator. Because of the 10-volt controlling voltage, 10 amperes of current will flow through the 1-ohm field winding. This draws 100 watts of input power (P = IE).

V\A
LOAD (10,000 WATTS) 115 VOLTS

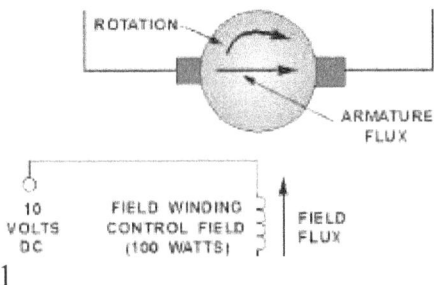

Figure 1-22.—Ordinary dc generator.

Assume that the characteristics of this generator enable it to produce approximately 87 amperes of armature current at 115 volts at the output terminals. This represents an output power of approximately 10,000 watts (P = IE). You can see that the power gain of this generator is 100. In effect, 100 watts controls 10,000 watts.

An amplidyne is a special type of dc generator. The following changes, for explanation purposes, will convert the typical dc generator above into an amplidyne.

The first step is to short the brushes together, as shown in figure 1-23. This removes nearly all of the resistance in the armature circuit.

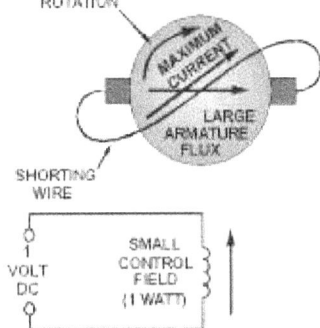

Figure 1-23.—Brushes shorted in a dc generator.

Because of the very low resistance in the armature circuit, a much lower control-field flux produces full-load armature current (full-load current in the armature is still about 87 amperes). The smaller control

field now requires a control voltage of only 1 volt and an input power of 1 watt (1 volt across 1 ohm causes 1 ampere of current, which produces 1 watt of input power).

The next step is to add another set of brushes. These now become the output brushes of the amplidyne. They are placed against the commutator in a position perpendicular to the original brushes, as shown in figure 1-24. The previously shorted brushes are now called the "quadrature" brushes. This is because they are in quadrature (perpendicular) to the output brushes. The output brushes are in line with the armature flux. Therefore, they pick off the voltage induced in the armature windings at this point. The voltage at the output will be the same as in the original generator, 115 volts in our example.

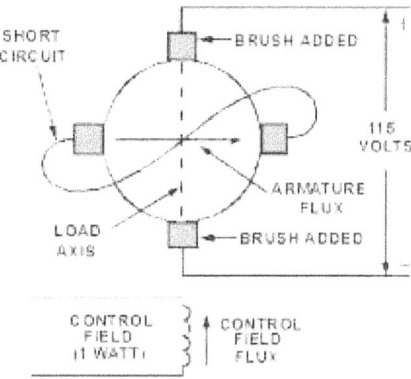

Figure 1-24.—Amplidyne load brushes.

As you have seen, the original generator produced a 10,000-watt output with a 100-watt input. The amplidyne produces the same 10,000-watt output with only a 1-watt input. This represents a gain of 10,000. The gain of the original generator has been greatly increased.

As previously stated, an amplidyne is used to provide large dc currents. The primary use of an amplidyne is in the positioning of heavy loads through the use of synchro/servo systems. Synchro/servo systems will be studied in a later module.

Assume that a very large turning force is required to rotate a heavy object, such as an antenna, to a very precise position. A low-power, relatively weak voltage representing the amount of antenna rotation required can be used to control the field winding of an amplidyne. Because of the amplidyne's ability to amplify, its output can be used to drive a powerful motor, which turns the heavy object (antenna). When the source of the input voltage senses the correct movement of the object, it drops the voltage to zero. The field is no longer strong enough to allow an output voltage to be developed, so the motor ceases to drive the object (antenna).

The above is an oversimplification and is not meant to describe a functioning system. The intent is to show a typical sequence of events between the demand for movement and the movement itself. It is meant to strengthen the idea that with the amplidyne, something large and heavy can be controlled very precisely by something very small, almost insignificant.

Q22. What is the purpose of a dc generator that has been modified to function as an amplidyne?

Q23. What is the formula used to determine the gain of an amplifying device?

Q24. What are the two inputs to an amplidyne?

SAFETY PRECAUTIONS

You must always observe safety precautions when working around electrical equipment to avoid injury to personnel and damage to equipment. Electrical equipment frequently has accessories that require separate sources of power. Lighting fixtures, heaters, externally powered temperature detectors, and alarm systems are examples of accessories whose terminals must be deenergized. When working on dc generators, you must check to ensure that all such circuits have been de-energized and tagged before you attempt any maintenance or repair work. You must also use the greatest care when working on or near the output terminals of dc generators.

SUMMARY

This chapter introduced you to the basic principles concerning direct current generators. The different types of dc generators and their characteristics were covered. The following information provides a summary of the major subjects of the chapter for your review.

MAGNETIC INDUCTION takes place when a conductor is moved in a magnetic field in such a way that it cuts flux lines, and a voltage (emf) is induced in the conductor.

THE LEFT-HAND RULE FOR GENERATORS states that when the thumb, forefinger, and middle finger of the left hand are extended at right angles to each other so that the thumb indicates the direction of movement of the conductor in the magnetic field, and the forefinger points in the direction of the flux lines (north to south), the middle finger shows the direction of induced emf in the conductor.

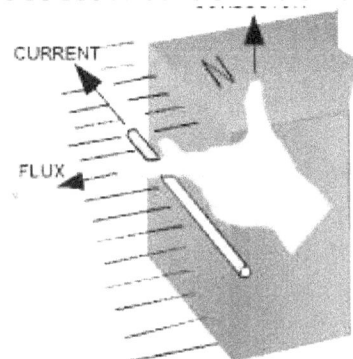

AN ELEMENTARY GENERATOR consists of a single coil rotated in a magnetic field. It produces an ac voltage.

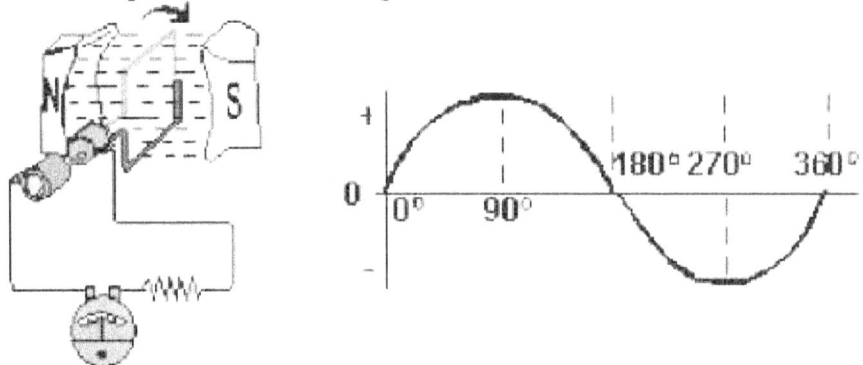

A BASIC DC GENERATOR results when you replace the slip rings of an elementary generator with a two-piece commutator, changing the output voltage to pulsating dc.

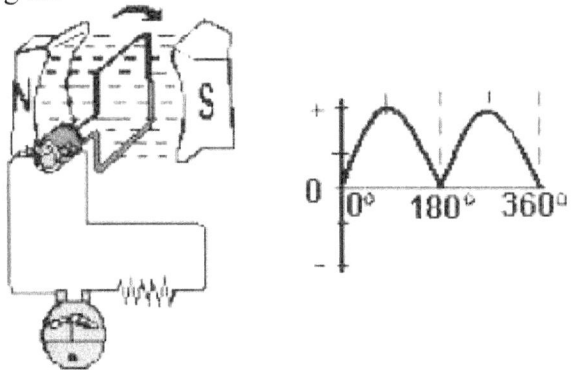

A MULTIPLE COIL ARMATURE (adding coils to the armature) decreases the ripple voltage in the output of a dc generator, and increases the output voltage.

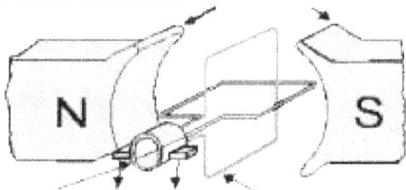

POLE PIECES
COMMUTATOR TWO-COIL ARMATURE
GENERATOR TERMINAL VOLTAGE

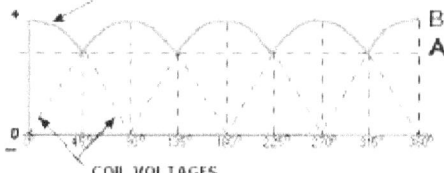

COIL VOLTAGES

A MULTIPOLE GENERATOR is the result of adding more field poles to a dc generator. They have much the same effect as adding coils to the armature. In practical generators, the poles are electromagnets.

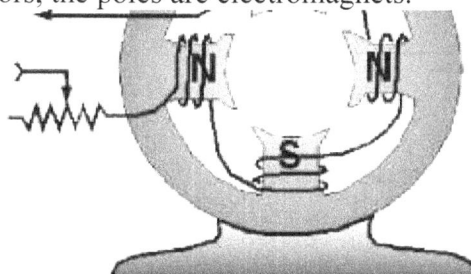

COMMUTATION is the process used to get direct current from a generator. The coil connections to the load must be reversed as the coil passes through the neutral plane. The brushes must be positioned so that commutation is accomplished without brush sparking.

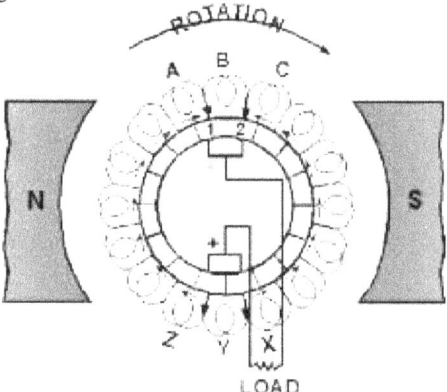

LOAD

ARMATURE REACTION takes place when armature current causes the armature to become an electromagnet. The armature field disturbs the field from the pole pieces. This results in a shift of the neutral plane in the direction of rotation.

OLD
M PLfiME L "™ hLHHh .NEUTRAL
PLANE
ROTATION

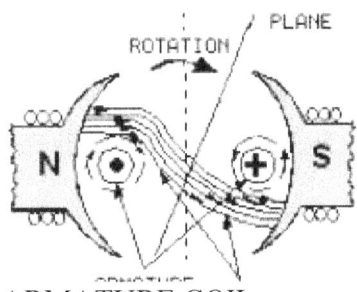

ARMATURE COIL
MAGNETIC FIELD RESULTING FROM INTERACTION

COMPENSATING WINDINGS AND INTERPOLES are used to counteract the effects of armature reaction. They are supplied by armature current and shift the neutral plane back to its original position.

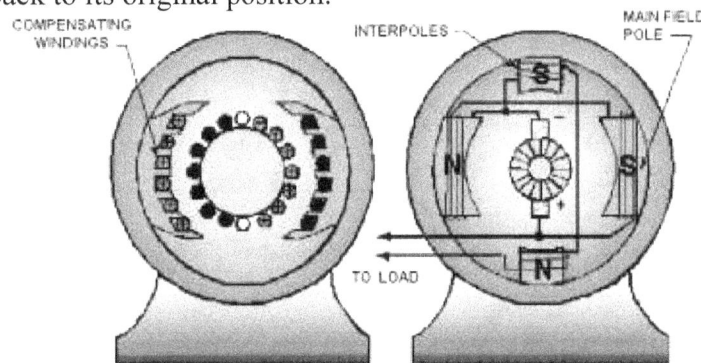

MOTOR REACTION is caused by the magnetic field that is set up in the armature. It tends to oppose the rotation of the armature, due to the attraction and repulsion forces between the armature field and the main field.

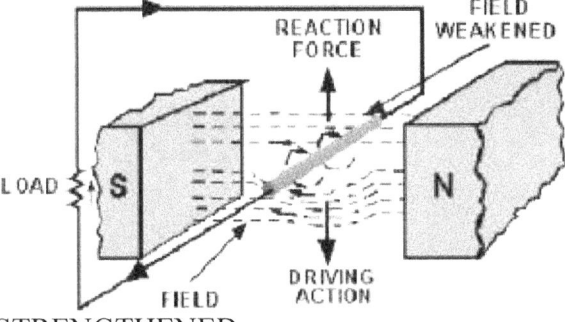

STRENGTHENED

ARMATURE LOSSES in dc generator armatures affect the outputs. These losses are as follows:

1. Copper losses are simply I^2R (heat) losses caused by current flowing through the resistance of the armature windings.

2. Eddy currents are induced in core material and cause heat.

3. Hysteresis losses occur due to the rapidly changing magnetic fields in the armature, resulting in heat.

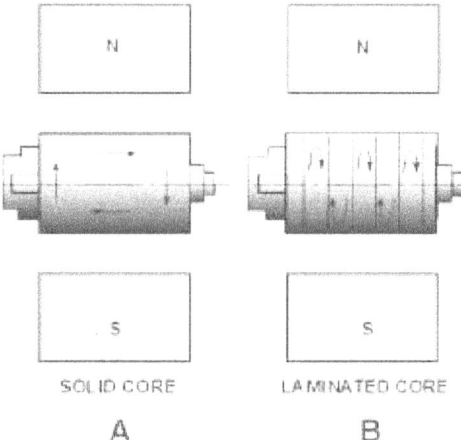

ARMATURE TYPES used in dc generators are the Gramme-ring (seldom used) and the drum-type, used in most applications.

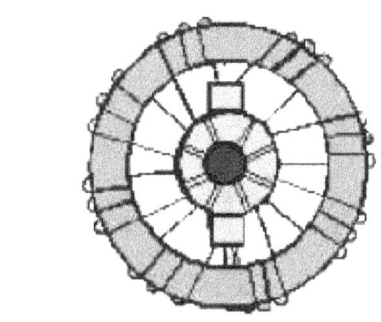

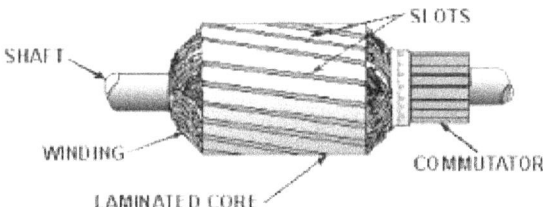

FIELD EXCITATION is the voltage applied to the main field windings. The current in the field coils determines the strength and the direction of the magnetic field.

SEPARATELY EXCITED GENERATORS receive current for field coils from an outside source such, as a battery or another dc generator.

SELF-EXCITED GENERATORS use their own output voltage to energize field coils.

SERIES-WOUND DC GENERATORS have field windings and armature windings connected in series. Outputs vary directly with load currents. Series-wound generators have few practical applications.

TO LOAD -* CIRCUIT
o-
SERIES FIELD
GENERATOR OUTPUT
ARMATURE*

SHUNT-WOUND DC GENERATORS have field windings and armature windings connected in parallel (shunt). The output varies inversely with load current.

GENERATOR OUTPUT
O—
SHUNT FIELD
6
/
ARMATURE

COMPOUND-WOUND DC GENERATORS have both series field windings and shunt field windings. These generators combine the characteristics of series and shunt generators. The output voltage remains relatively constant for all values of load current within the design of the generator. Compound generators are used in many applications because of the relatively constant voltage.

AMPLIDYNES are dc generators that are designed to act as high-gain amplifiers. By short-circuiting the brushes in a normal dc generator and adding another set of brushes perpendicular to the original ones, an amplidyne is formed. Its power output may be up to 10,000 times larger than the power input to its control windings.

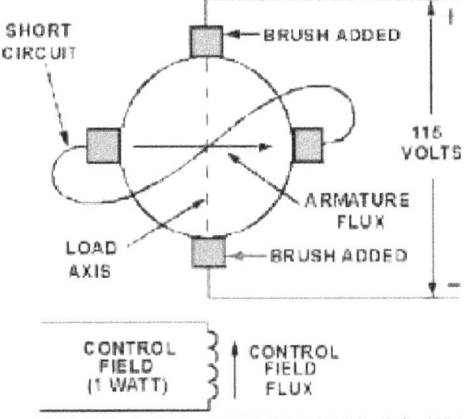

ANSWERS TO QUESTIONS Q1. THROUGH Q24.

A1. Magnetic induction.

A2. The left-hand rule for generators.

A3. To conduct the currents induced in the armature to an external load.

A4. No flux lines are cut.

A5. A commutator

A6. The point at which the voltage is zero across the two segments.

A7. Two.

A8. Four

A9. By varying the input voltage to the field coils.

A10. Improper commutation.

A11. Distortion of the main field due to the effects of armature current.

A12. To counter act armature reaction.

A13. A force which causes opposition to applied turning force.

A14. Resistance in the armature coils, which increases with temperature.

A15. By laminating the core material.

A16. Drum-type armatures are more efficient, because flux lines are cut by both sides of each coil.

A17. Higher load currents are possible.
A18. Series-wound, shunt-wound, and compound-wound.
A19. Output voltage varies as the load varies.
A20. Voltage regulation.
A21. Parallel operation.
A22. It can serve as a power amplifier.
A23. Gain = output + input.
A24. The mechanical force applied to turn the amplidyne, and the electrical input signal.

CHAPTER 2
DIRECT CURRENT MOTORS

LEARNING OBJECTIVES

Upon completion of this chapter you will be able to:

1. State the factors that determine the direction of rotation in a dc motor.
2. State the right-hand rule for motors.
3. Describe the main differences and similarities between a dc generator and a dc motor.
4. Describe the cause and effect of counter emf in a dc motor.
5. Explain the term "load" as it pertains to an electric motor.
6. List the advantages and disadvantages of the different types of dc motors.
7. Compare the types of armatures and uses for each.
8. Discuss the means of controlling the speed and direction of a dc motor.
9. Describe the effect of armature reaction in a dc motor. 10. Explain the need for a starting resistor in a dc motor.

INTRODUCTION

The dc motor is a mechanical workhorse, that can be used in many different ways. Many large pieces of equipment depend on a dc motor for their power to move. The speed and direction of rotation of a dc motor are easily controlled. This makes it especially useful for operating equipment, such as winches, cranes, and missile launchers, which must move in different directions and at varying speeds.

PRINCIPLES OF OPERATION

The operation of a dc motor is based on the following principle:

A current-carrying conductor placed in a magnetic field, perpendicular to the lines of flux, tends to move in a direction perpendicular to the magnetic lines of flux.

There is a definite relationship between the direction of the magnetic field, the direction of current in the conductor, and the direction in which the conductor tends to move. This relationship is best explained by using the RIGHT-HAND RULE FOR MOTORS (fig. 2-1).

MOTION OF

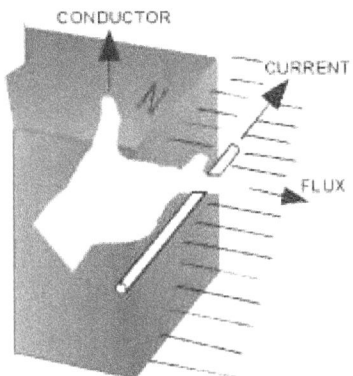

Figure 2-1.—Right-hand rule for motors.

To find the direction of motion of a conductor, extend the thumb, forefinger, and middle finger of your right hand so they are at right angles to each other. If the forefinger is pointed in the direction of magnetic flux (north to south) and the middle finger is pointed in the direction of current flow in the conductor, the thumb will point in the direction the conductor will move.

Stated very simply, a dc motor rotates as a result of two magnetic fields interacting with each other. The armature of a dc motor acts like an electromagnet when current flows through its coils. Since the armature is located within the magnetic field of the field poles, these two magnetic fields interact. Like magnetic poles repel each other, and unlike magnetic poles attract each other. As in the dc generator, the dc motor has field poles that are stationary and an armature that turns on bearings in the space between the field poles. The armature of a dc motor has windings on it just like the armature of a dc generator. These windings are also connected to commutator segments. A dc motor consists of the same components as a dc generator. In fact, most dc generators can be made to act as motors, and vice versa.

Look at the simple dc motor shown in figure 2-2. It has two field poles, one a north pole and one a south pole. The magnetic lines of force extend across the opening between the poles from north to south.

DIRECTION OF ROTATION

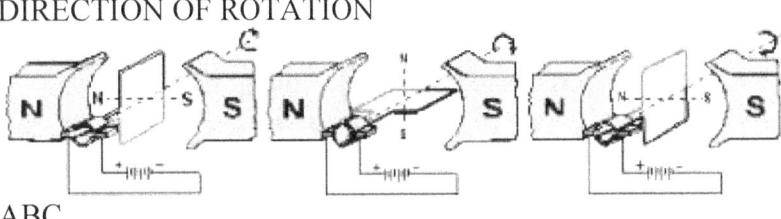

ABC

Figure 2-2.—Dc motor armature rotation.

The armature in this simple dc motor is a single loop of wire, just as in the simple armature you studied at the beginning of the chapter on dc generators. The loop of wire in the dc motor, however, has

current flowing through it from an external source. This current causes a magnetic field to be produced. This field is indicated by the dotted line through the loops. The loop (armature) field is both attracted and repelled by the field from the field poles. Since the current through the loop goes around in the direction of the arrows, the north pole of the armature is at the upper left, and the south pole of the armature is at the lower right, as shown in figure 2-2, (view A). Of course, as the loop (armature) turns, these magnetic poles turn with it. Now, as shown in the illustrations, the north armature pole is repelled from the north field pole and attracted to the right by the south field pole. Likewise, the south armature pole is repelled from the south field pole and is attracted to the left by the north field pole. This action causes the armature to turn in a clockwise direction, as shown in figure 2-2 (view B).

After the loop has turned far enough so that its north pole is exactly opposite the south field pole, the brushes advance to the next segments. This changes the direction of current flow through the armature loop. Also, it changes the polarity of the armature field, as shown in figure 2-2 (view C). The magnetic fields again repel and attract each other, and the armature continues to turn.

In this simple motor, the momentum of the rotating armature carries the armature past the position where the unlike poles are exactly lined up. However, if these fields are exactly lined up when the armature current is turned on, there is no momentum to start the armature moving. In this case, the motor would not rotate. It would be necessary to give a motor like this a spin to start it. This disadvantage does not exist when there are more turns on the armature, because there is more than one armature field. No two armature fields could be exactly aligned with the field from the field poles at the same time.

Q1. What factors determine the direction of rotation in a dc motor?

Q2. The right-hand rule for motors is used to find the relationship between what motor characteristics?

Q3. What are the differences between the components of a dc generator and a dc motor?

COUNTER EMF

While a dc motor is running, it acts somewhat like a dc generator. There is a magnetic field from the field poles, and a loop of wire is turning and cutting this magnetic field. For the moment, disregard the fact that there is current flowing through the loop of wire from the battery. As the loop sides cut the magnetic field, a voltage is induced in them, the same as it was in the loop sides of the dc generator. This induced voltage causes current to flow in the loop.

Now, consider the relative direction between this current and the current that causes the motor to run. First, check the direction the current flows as a result of the generator action taking place (view A of fig. 2-2). (Apply the left-hand rule for generators which was discussed in the last chapter.) Using the left hand, hold it so that the forefinger points in the direction of the magnetic field (north to south) and the thumb points in the direction that the black side of the armature moves (up). Your middle finger then points out of the paper (toward you), showing the direction of current flow caused by the generator action in the black half of the armature. This is in the direction opposite to that of the battery current. Since this generator-action voltage is opposite that of the battery, it is called "counter emf." (The letters emf stand for electromotive force, which is another name for voltage.) The two currents are flowing in opposite directions. This proves that the battery voltage and the counter emf are opposite in polarity.

At the beginning of this discussion, we disregarded armature current while explaining how counter emf was generated. Then, we showed that normal armature current flowed opposite to the current created by the counter emf. We talked about two opposite currents that flow at the same time. However, this is a
bit oversimplified, as you may already suspect. Actually, only one current flows.

Because the counter emf can never become as large as the applied voltage, and because they are of opposite polarity as we have seen, the counter emf effectively cancels part of the armature voltage. The single current that flows is armature current, but it is greatly reduced because of the counter emf.

In a dc motor, there is always a counter emf developed. This counter emf cannot be equal to or greater than the applied battery voltage; if it were, the motor would not run. The counter emf is always a little less. However, the counter emf opposes the applied voltage enough to keep the armature current from the battery to a fairly low value. If there were no such thing as counter emf, much more current would flow through the armature, and the motor would run much faster. However, there is no way to avoid the counter emf.

Q4. What causes counter emf in a dc motor?

Q5. What motor characteristic is affected by counter emf?

MOTOR LOADS

Motors are used to turn mechanical devices, such as water pumps, grinding wheels, fan blades, and circular saws. For example, when a motor is turning a water pump, the water pump is the load. The water pump is the mechanical device that the motor must move. This is the definition of a motor load.

As with electrical loads, the mechanical load connected to a dc motor affects many electrical quantities. Such things as the power drawn from the line, amount of current, speed, efficiency, etc., are all partially controlled by the size of the load. The physical and electrical characteristics of the motor must be matched to the requirements of the load if the work is to be done without the possibility of damage to either the load or the motor.

Q6. What is the load on a dc motor?

PRACTICAL DC MOTORS

As you have seen, dc motors are electrically identical to dc generators. In fact, the same dc machine may be driven mechanically to generate a voltage, or it may be driven electrically to move a mechanical load. While this is not normally done, it does point out the similarities between the two machines. These similarities will be used in the remainder of this chapter to introduce you to practical dc motors. You will immediately recognize series, shunt, and compound types of motors as being directly related to their generator counterparts.

SERIES DC MOTOR

In a series dc motor, the field is connected in series with the armature. The field is wound with a few turns of large wire, because it must carry full armature current. The circuit for a series dc motor is shown in figure 2-3.

SERIES FIELD

INPUT VOLTAGE

O—

ARMATURE-

Figure 2-3.—Series-wound dc motor.

This type of motor develops a very large amount of turning force, called torque, from a standstill. Because of this characteristic, the series dc motor can be used to operate small electric appliances, portable electric tools, cranes, winches, hoists, and the like.

Another characteristic is that the speed varies widely between no-load and full-load. Series motors cannot be used where a relatively constant speed is required under conditions of varying load.

A major disadvantage of the series motor is related to the speed characteristic mentioned in the last paragraph. The speed of a series motor with no load connected to it increases to the point where the motor may become damaged. Usually, either the bearings are damaged or the windings fly out of the slots in the armature. There is a danger to both equipment and personnel. Some load must ALWAYS be connected to a series motor before you turn it on. This precaution is primarily for large motors. Small motors, such as those used in electric hand drills, have enough internal friction to load themselves.

A final advantage of series motors is that they can be operated by using either an ac or dc power source. This will be covered in the chapter on ac motors.

Q7. What is the main disadvantage of a series motor?

Q8. What is the main advantage of a series motor?

SHUNT MOTOR

A shunt motor is connected in the same way as a shunt generator. The field windings are connected in parallel (shunt) with the armature windings. The circuit for a shunt motor is shown in figure 2-4.

VOLTAGE
SHUNT FIELD
6
/
ARMATURE

Figure 2-4.—Shunt-wound dc motor.

Once you adjust the speed of a dc shunt motor, the speed remains relatively constant even under changing load conditions. One reason for this is that the field flux remains constant. A constant voltage across the field makes the field independent of variations in the armature circuit.

If the load on the motor is increased, the motor tends to slow down. When this happens, the counter emf generated in the armature decreases. This causes a corresponding decrease in the opposition to battery current flow through the armature. Armature current increases, causing the motor to speed up. The conditions that established the original speed are reestablished, and the original speed is maintained.

Conversely, if the motor load is decreased, the motor tends to increase speed; counter emf increases, armature current decreases, and the speed decreases.

In each case, all of this happens so rapidly that any actual change in speed is slight. There is instantaneous tendency to change rather than a large fluctuation in speed.

Q9. What advantage does a shunt motor have over a series motor?

COMPOUND MOTOR

A compound motor has two field windings, as shown in figure 2-5. One is a shunt field connected in parallel with the armature; the other is a series field that is connected in series with the armature. The shunt field gives this type of motor the constant speed advantage of a regular shunt motor. The series field gives it the advantage of being able to develop a large torque when the motor is started under a heavy load. It should not be a surprise that the compound motor has both shunt- and series-motor characteristics.

INPUT

VOLTAGE O-i
SERIES I FIELD
SHUNT
FIELD
LONG
ARMATURE
O— 1 INPUT
VOLTAGE O—t
SERIES FIELD
SHUNT FIELD SHORT
ARMATURE
(A) LONG SHUNT (B) SHORT SHUNT

Figure 2-5.—Compound-wound dc motor.

When the shunt field is connected in parallel with the series field and armature, it is called a "long shunt" as shown in figure 2-5, (view A). Otherwise, it is called a "short shunt", as shown in figure 2-5, (view B).

TYPES OF ARMATURES

As with dc generators, dc motors can be constructed using one of two types of armatures. A brief review of the Gramme-ring and drum-wound armatures is necessary to emphasize the similarities between dc generators and dc motors.

GRAMME-RING ARMATURE

The Gramme-ring armature is constructed by winding an insulated wire around a soft-iron ring (fig. 2-6). Eight equally spaced connections are made to the winding. Each of these is connected to a commutator segment. The brushes touch only the top and bottom segments. There are two parallel paths for current to follow — one up the left side and one up the right side. These paths are completed through the top brush back to the positive lead of the battery.

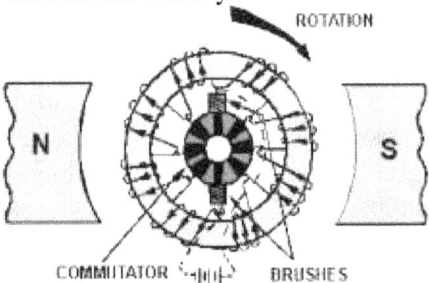

Figure 2-6.—Gramme-ring armature.

To check the direction of rotation of this armature, you should use the right-hand rule for motors. Hold your thumb, forefinger, and middle finger at right angles. Point your forefinger in the direction of field flux; in this case, from left to right. Now turn your wrist so that your middle finger points in the direction that the current flows in the winding on the outside of the ring. Note that current flows into the page (away from you) in the left-hand windings and out of the page (toward you) in the right-hand windings. Your thumb now points in the direction that the winding will move.

The Gramme-ring armature is seldom used in modem dc motors. The windings on the inside of the ring are shielded from magnetic flux, which causes this type of armature to be inefficient. The Gramme-ring armature is discussed primarily to help you better understand the drum-wound armature.

DRUM-WOUND ARMATURE

The drum-wound armature is generally used in ac motors. It is identical to the drum winding discussed in the chapter on dc generators.

If the drum-wound armature were cut in half, an end view at the cut would resemble the drawing in figure 2-7, (view A), Figure 2-7, (view B) is a side view of the armature and pole pieces. Notice that the length of each conductor is positioned parallel to the faces of the pole pieces. Therefore, each conductor of the armature can cut the maximum flux of the motor field. The inefficiency of the Gramme-ring armature is overcome by this positioning.

ARMATURE
COIL (IH SLOT)
A
END VIEW (CROSS SECTION)
B
SIDE VIEW
ARMATURE CORE

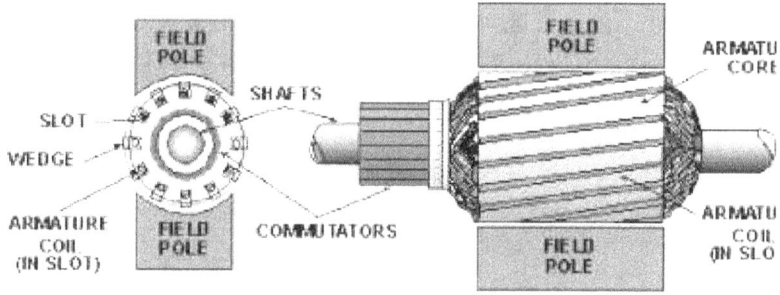

ARMATURE
COIL (IH SLOT)

Figure 2-7.—Drum-type armature.

The direction of current flow is marked in each conductor in figure 2-7, (view A) as though the armature were turning in a magnetic field. The dots show that current is flowing toward you on the left side, and the crosses show that the current is flowing away from you on the right side.

Strips of insulation are inserted in the slots to keep windings in place when the armature spins. These are shown as wedges in figure 2-7, (view A).

Q10. Why is the Gramme-ring armature not more widely used?

Q11. How is the disadvantage of the Gramme-ring armature overcome in the drum-wound armature?

DIRECTION OF ROTATION

The direction of rotation of a dc motor depends on the direction of the magnetic field and the direction of current flow in the armature. If either the direction of the field or the direction of current flow through the armature is reversed, the rotation of the motor will reverse. However, if both of these factors are reversed at the same time, the motor will continue rotating in the same direction. In actual practice, the field excitation voltage is reversed in order to reverse motor direction.

Ordinarily, a motor is set up to do a particular job that requires a fixed direction of rotation. However, there are times when it is necessary to change the direction of rotation, such as a drive motor for a gun turret or missile launcher. Each of these must be able to move in both directions. Remember, the connections of either the armature or the field must be reversed, but not both. In such applications, the proper connections are brought out to a reversing switch.

Q12. In a dc motor that must be able to rotate in both directions, how is the direction changed?

A motor whose speed can be controlled is called a variable-speed motor; dc motors are variable-speed motors. The speed of a dc motor is changed by changing the current in the field or by changing the current in the armature.

When the field current is decreased, the field flux is reduced, and the counter emf decreases. This permits more armature current. Therefore, the motor speeds up. When the field current is increased, the field flux is increased. More counter emf is developed, which opposes the armature current. The armature current then decreases, and the motor slows down.

When the voltage applied to the armature is decreased, the armature current is decreased, and the motor again slows down. When the armature voltage and current are both increased, the motor speeds up.

In a shunt motor, speed is usually controlled by a rheostat connected in series with the field windings, as shown in figure 2-8. When the resistance of the rheostat is increased, the current through the field winding is decreased. The decreased flux momentarily decreases the counter emf. The motor then speeds up, and the increase in counter emf keeps the armature current the same. In a similar manner, a decrease in rheostat resistance increases the current flow through the field windings and causes the motor to slow down.

MOTOR SPEED

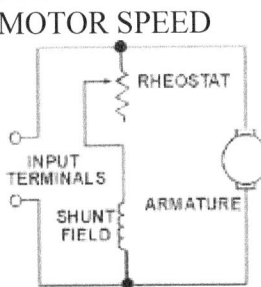

Figure 2-8.—Controlling motor speed.

In a series motor, the rheostat speed control may be connected either in parallel or in series with the armature windings. In either case, moving the rheostat in a direction that lowers the voltage across the armature lowers the current through the armature and slows the motor. Moving the rheostat in a direction that increases the voltage and current through the armature increases motor speed.

Q13. What is the effect on motor speed if the field current is increased?

ARMATURE REACTION

You will remember that the subject of armature reaction was covered in the previous chapter on dc generators. The reasons for armature reaction and the methods of compensating for its effects are basically the same for dc motors as for dc generators.

Figure 2-9 reiterates for you the distorting effect that the armature field has on the flux between the pole pieces. Notice, however, that the effect has shifted the neutral plane backward, against the direction of rotation. This is different from a dc generator, where the neutral plane shifted forward in the direction of rotation.

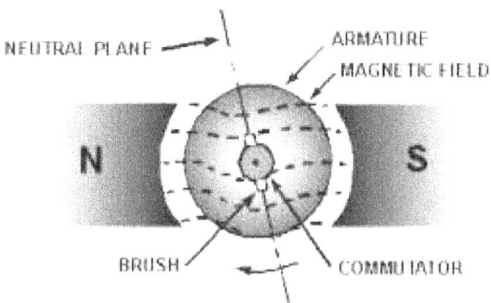

Figure 2-9.—Armature reaction.

As before, the brushes must be shifted to the new neutral plane. As shown in figure 2-9, the shift is counterclockwise. Again, the proper location is reached when there is no sparking from the brushes.

Q14. Armature reaction in a dc motor causes a shift of the neutral plane in which direction?

Compensating windings and interpoles, two more "old" subjects, cancel armature reaction in dc motors. Shifting brushes reduces sparking, but it also makes the field less effective. Canceling armature reaction eliminates the need to shift brushes in the first place.

Compensating windings and interpoles are as important in motors as they are in generators. Compensating windings are relatively expensive; therefore, most large dc motors depend on interpoles to correct armature reaction. Compensating windings are the same in motors as they are in generators. Interpoles, however, are slightly different. The difference is that in a generator the interpole has the same polarity as the main pole AHEAD of it in the direction of rotation. In a motor the interpole has the same polarity as the main pole FOLLOWING it.

The interpole coil in a motor is connected to carry the armature current the same as in a generator. As the load varies, the interpole flux varies, and commutation is automatically corrected as the load changes. It is not necessary to shift the brushes when there is an increase or decrease in load. The brushes are located on the no-load neutral plane. They remain in that position for all conditions of load.

Q15. What current flows in the interpole windings?

The dc motor is reversed by reversing the direction of the current in the armature. When the armature current is reversed, the current through the interpole is also reversed. Therefore, the interpole still has the proper polarity to provide automatic commutation.

MANUAL AND AUTOMATIC STARTERS

Because the dc resistance of most motor armatures is low (0.05 to 0.5 ohm), and because the counter emf does not exist until the armature begins to turn, it is necessary to use an external starting resistance in series with the armature of a dc motor to keep the initial armature current to a safe value. As the armature begins to turn, counter emf increases; and, since the counter emf opposes the applied voltage, the armature current is reduced. The external resistance in series with the armature is decreased or eliminated as the motor comes up to normal speed and full voltage is applied across the armature.

Controlling the starting resistance in a dc motor is accomplished either manually, by an operator, or by any of several automatic devices. The automatic devices are usually just switches controlled by motor speed sensors. Automatic starters are not covered in detail in this module.

Q16. What is the purpose of starting resistors?

SUMMARY

This chapter presented the operating principles and characteristics of direct-current motors. The following information provides a summary of the main subjects for review.

The main PRINCIPLE OF A DC MOTOR is that current flow through the armature coil causes the armature to act as a magnet. The armature poles are attracted to field poles of opposite polarity, causing the armature to rotate.

The CONSTRUCTION of a dc motor is almost identical to that of a dc generator, both physically and electrically. In fact, most dc generators can be made to act as dc motors, and vice versa.

COMMUTATION IN A DC MOTOR is the process of reversing armature current at the moment when unlike poles of the armature and field are facing each other, thereby reversing the polarity of the armature field. Like poles of the armature and field then repel each other, causing armature rotation to continue.

DIRECTION OF ROTATION

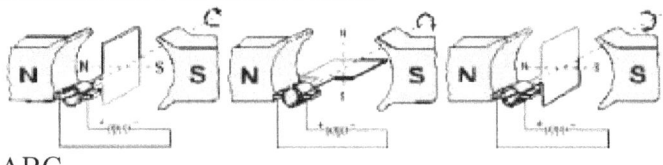

ABC

COUNTER-ELECTROMOTIVE FORCE is generated in a dc motor as armature coils cut the field flux. This emf opposes the applied voltage, and limits the flow of armature current.

In SERIES MOTORS, the field windings are connected in series with the armature coil. The field strength varies with changes in armature current. When its speed is reduced by a load, the series motor develops greater torque. Its starting torque is greater than other types of dc motors. Its speed varies widely between full-load and no-load. Unloaded operation of large machines is dangerous.

SERIES FIELD
INPUT VOLTAGE
O—
ARMATURE*

In SHUNT MOTORS, the field windings are connected in parallel (shunt) across the armature coil. The field strength is independent of the armature current. Shunt-motor speed varies only slightly with changes in load, and the starting torque is less than that of other types of dc motors.

INPUT
VOLTAGE
SHUNT FIELD
ARMATURE

In COMPOUND MOTORS, one set of field windings is connected in series with the armature, and one set is connected in parallel. The speed and torque characteristics are a combination of the desirable characteristics of both series and shunt motors.

LOAD on a motor is the physical object to be moved by the motor.

DC MOTOR ARMATURES are of two types. They are the Gramme-ring and the drum-wound types.

THE GRAMME-RING ARMATURE is inefficient since part of each armature coil is prevented from cutting flux lines. Gramme-ring wound armatures are seldom used for this reason.

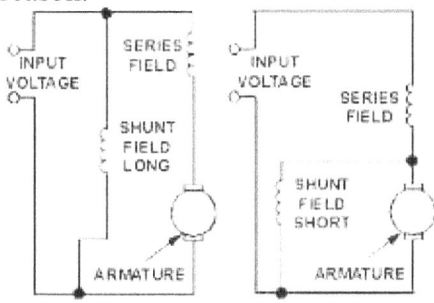

(A) LONG SHUNT
(B) SHORT SHUNT
2-13

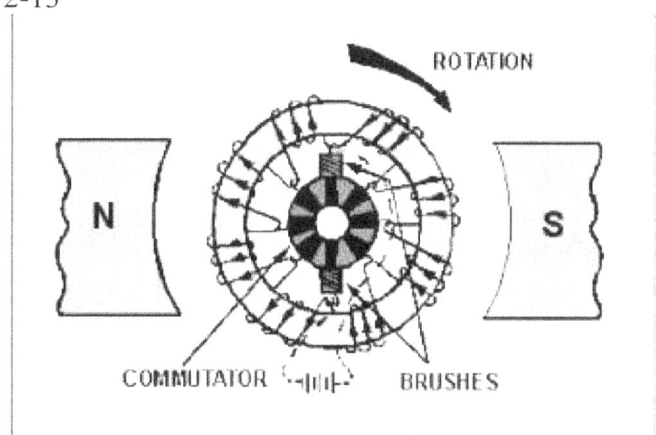

THE DRUM-WOUND ARMATURE consists of coils actually wound around the armature core so that all coil surfaces are exposed to the magnetic field. Nearly all dc motors have drum-wound armatures.

ARMATURE
COIL (IN SLOT)
A
END VIEW (CROSS SECTION)
B
SIDE VIEW
ARMATURE CORE

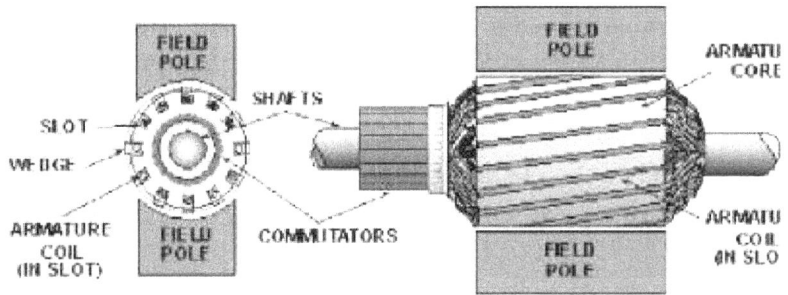

ARMATURE
COIL fIH SLOT)

MOTOR REVERSAL in a dc motor can be accomplished by reversing the field connections or by reversing the armature connections. If both are reversed, rotation will continue in the original direction.

SPEED CONTROL IN A DC MOTOR is maintained by varying the resistance either in series with the field coil or in series with the armature coil. Increasing shunt-field circuit resistance increases motor speed. Increasing the armature circuit resistance decreases motor speed.

VOLTAGE
SHUNT FIELD
6
ARMATURE

ARMATURE REACTION is the distortion of the main field in a motor by the armature field. This causes the neutral plane to be shifted in the direction opposite to that of armature rotation. Interpoles and compensating windings are used to reduce the effect of armature reaction on motor operation.

STARTING RESISTORS are necessary since the dc resistance of a motor armature is very low. Excessive current will flow when dc voltage is first applied unless current is limited in some way. Adding resistance in series with the armature windings reduces initial current. It may then be removed after counter emf has been built up.

ANSWERS TO QUESTIONS Ql. THROUGH Q16.

A1. Direction of armature current, and direction of magnetic flux in field.

A2. Direction of conductor movement (rotation), direction offlux, and the direction of current flow. A3. There are no differences. A4. Generator action.

A5. Speed.

A6. The device to be driven by the motor.

A 7. It must have a load connected to avoid damage from excess speed.

A8. High torque (turningforce) at low speed.

A9. It maintains a constant speed under varying loads.

A10. Only outside of coils cut flux (inefficient).

All. By winding the armature in a way that places the entire coil where it is exposed to maximum flux.

A12. By reversing either field or armature connections.

2-15

A13. Motor will slow down.

A14. Opposite the rotation.

A15. Armature current.

A16. To limit armature current until counter emf builds up.

CHAPTER 3
ALTERNATING CURRENT GENERATORS
LEARNING OBJECTIVES

Upon completion of this chapter, you will be able to:

1. Describe the principle of magnetic induction as it applies to ac generators.
2. Describe the differences between the two basic types of ac generators.
3. List the advantages and disadvantages of the two types of ac generators.
4. Describe exciter generators within alternators; discuss construction and purpose.
5. Compare the types of rotors used in ac generators, and the applications of each type to different prime movers.
6. Explain the factors that determine the maximum power output of an ac generator, and the effect of these factors in rating generators.
7. Explain the operation of multiphase ac generators and compare with single-phase.
8. Describe the relationships between the individual output and resultant vectorial sum voltages in multiphase generators.
9. Explain, using diagrams, the different methods of connecting three-phase alternators and transformers.
10. List the factors that determine the frequency and voltage of the alternator output.
11. Explain the terms voltage control and voltage regulation in ac generators, and list the factors that affect each quantity.
12. Describe the purpose and procedure of parallel generator operation.

INTRODUCTION

Most of the electrical power used aboard Navy ships and aircraft as well as in civilian applications is ac. As a result, the ac generator is the most important means of producing electrical power. Ac generators, generally called alternators, vary greatly in size depending upon the load to which they supply power. For example, the alternators in use at hydroelectric plants, such as Hoover Dam, are tremendous in size, generating thousands of kilowatts at very high voltage levels. Another example is the alternator in a typical automobile, which is very small by comparison. It weighs only a few pounds and produces between 100 and 200 watts of power, usually at a potential of 12 volts.

Many of the terms and principles covered in this chapter will be familiar to you. They are the same as those covered in the chapter on dc generators. You are encouraged to refer back, as needed, and to refer

to any other source that will help you master the subject of this chapter. No one source meets the complete needs of everyone.

BASIC AC GENERATORS

Regardless of size, all electrical generators, whether dc or ac, depend upon the principle of magnetic induction. An emf is induced in a coil as a result of (1) a coil cutting through a magnetic field, or (2) a magnetic field cutting through a coil. As long as there is relative motion between a conductor and a magnetic field, a voltage will be induced in the conductor. That part of a generator that produces the magnetic field is called the field. That part in which the voltage is induced is called the armature. For relative motion to take place between the conductor and the magnetic field, all generators must have two mechanical parts — a rotor and a stator. The ROT or is the part that ROT ates; the STAT or is the part that remains STAT ionary. In a dc generator, the armature is always the rotor. In alternators, the armature may be either the rotor or stator.

Q1. Magnetic induction occurs when there is relative motion between what two elements?

ROTATING-ARMATURE ALTERNATORS

The rotating-armature alternator is similar in construction to the dc generator in that the armature rotates in a stationary magnetic field as shown in figure 3-1, view A. In the dc generator, the emf generated in the armature windings is converted from ac to dc by means of the commutator. In the alternator, the generated ac is brought to the load unchanged by means of slip rings. The rotating armature is found only in alternators of low power rating and generally is not used to supply electric power in large quantities.

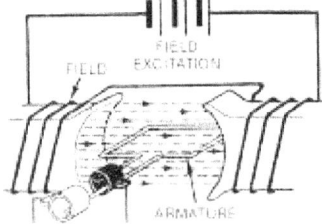

AC OUTPUT
A ROTATING ARMATURE ALTERNATOR

^AC OUTPUT^

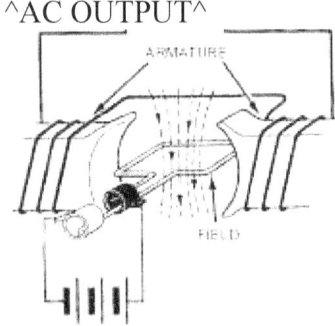

FIELD EXCITATION
B ROTATING FIELD ALTERNATOR

Figure 3-1.—Types of ac generators.

ROTATING-FIELD ALTERNATORS

The rotating-field alternator has a stationary armature winding and a rotating-field winding as shown in figure 3-1, view B The advantage of having a stationary armature winding is that the generated voltage can be connected directly to the load.

A rotating armature requires slip rings and brushes to conduct the current from the armature to the load. The armature, brushes, and slip rings are difficult to insulate, and arc-overs and short circuits can result at high voltages. For this reason, high-voltage alternators are usually of the rotating-field type. Since the voltage applied to the rotating field is low voltage dc, the problem of high voltage arc-over at the slip rings does not exist.

The stationary armature, or stator, of this type of alternator holds the windings that are cut by the rotating magnetic field. The voltage generated in the armature as a result of this cutting action is the ac power that will be applied to the load.

The stators of all rotating-field alternators are about the same. The stator consists of a laminated iron core with the armature windings embedded in this core as shown in figure 3-2. The core is secured to the stator frame.

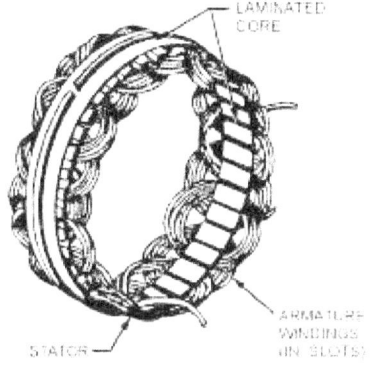

ASSEMBLY

Figure 3-2.—Stationary armature windings.

Q2. What is the part of an alternator in which the output voltage is generated?
Q3. What are the two basic types of alternators?
Q4. What is the main advantage of the rotating field alternator?

PRACTICAL ALTERNATORS

The alternators described so far in this chapter are ELEMENTARY in nature; they are seldom used except as examples to aid in understanding practical alternators.

The remainder of this chapter will relate the principles of the elementary alternator to the alternators actually in use in the civilian community, as well as aboard Navy ships and aircraft. The following paragraphs in this chapter will introduce such concepts as prime movers, field excitation, armature characteristics and limitations, single-phase and polyphase alternators, controls, regulation, and parallel operation.

FUNCTIONS OF ALTERNATOR COMPONENTS

A typical rotating-field ac generator consists of an alternator and a smaller dc generator built into a single unit. The output of the alternator section supplies alternating voltage to the load. The only purpose for the dc exciter generator is to supply the direct current required to maintain the alternator field. This dc generator is referred to as the exciter. A typical alternator is shown in figure 3-3, view A; figure 3-3, view B, is a simplified schematic of the generator.

A C FIELD INPUT A C FIELD WINDINGS
SLIP RING (5) (ROTOR) (6)

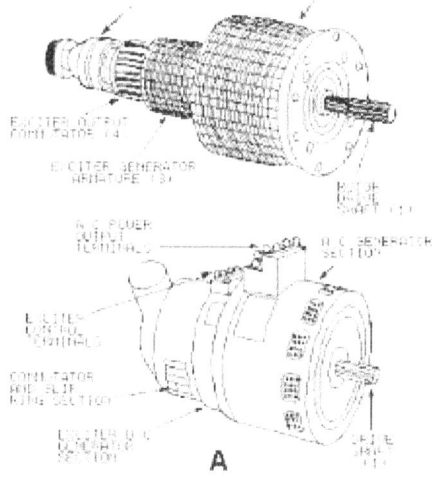

A

EXCITER CONTROL TERMINALS
fl C POWER OUTPUT TERMINALS

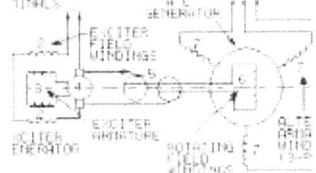

B

EXCITER GENERATOR
ALTERNATOR ARMATURE WINDINGS (3-PHASE)

Figure 3-3.—Ac generator pictorial and schematic drawings.

The exciter is a dc, shunt-wound, self-excited generator. The exciter shunt field (2) creates an area of intense magnetic flux between its poles. When the exciter armature (3) is rotated in the exciter-field flux, voltage is induced in the exciter armature windings. The output from the exciter commutator (4) is connected through brushes and slip rings (5) to the alternator field. Since this is direct current already converted by the exciter commutator, the current always flows in one direction through the alternator field (6). Thus, a fixed-polarity magnetic field is maintained at all times in the alternator field windings. When the alternator field is rotated, its magnetic flux is passed through and across the alternator armature windings (7).

The armature is wound for a three-phase output, which will be covered later in this chapter. Remember, a voltage is induced in a conductor if it is stationary and a magnetic field is passed across the conductor, the same as if the field is stationary and the conductor is moved. The alternating voltage in the ac generator armature windings is connected through fixed terminals to the ac load.

Q5. Most large alternators have a small dc generator built into them. What is its purpose ?

PRIME MOVERS

All generators, large and small, ac and dc, require a source of mechanical power to turn their rotors. This source of mechanical energy is called a prime mover.

Prime movers are divided into two classes for generators-high-speed and low-speed. Steam and gas turbines are high-speed prime movers, while internal-combustion engines, water, and electric motors are considered low-speed prime movers.

The type of prime mover plays an important part in the design of alternators since the speed at which the rotor is turned determines certain characteristics of alternator construction and operation.

ALTERNATOR ROTORS

There are two types of rotors used in rotating-field alternators. They are called the turbine-driven and salient-pole rotors.

As you may have guessed, the turbine-driven rotor shown in figure 3-4, view A, is used when the prime mover is a high-speed turbine. The windings in the turbine-driven rotor are arranged to form two or four distinct poles. The windings are firmly embedded in slots to withstand the tremendous centrifugal forces encountered at high speeds.

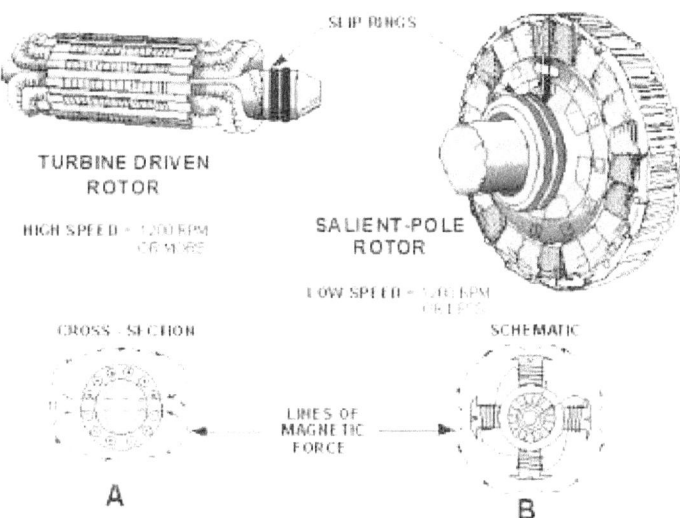

Figure 3-4.—Types of rotors used in alternators.

The salient-pole rotor shown in figure 3-4, view B, is used in low-speed alternators. The salient-pole rotor often consists of several separately wound pole pieces, bolted to the frame of the rotor.

If you could compare the physical size of the two types of rotors with the same electrical characteristics, you would see that the salient-pole rotor would have a greater diameter. At the same number of revolutions per minute, it has a greater centrifugal force than does the turbine-driven rotor. To

reduce this force to a safe level so that the windings will not be thrown out of the machine, the salient pole is used only in low-speed designs.

ALTERNATOR CHARACTERISTICS AND LIMITATIONS

Alternators are rated according to the voltage they are designed to produce and the maximum current they are capable of providing. The maximum current that can be supplied by an alternator depends upon the maximum heating loss that can be sustained in the armature. This heating loss (which is an I^2R power loss) acts to heat the conductors, and if excessive, destroys the insulation. Thus, alternators are rated in terms of this current and in terms of the voltage output — the alternator rating in small units is in volt-amperes; in large units it is kilovolt-amperes.

When an alternator leaves the factory, it is already destined to do a very specific job. The speed at which it is designed to rotate, the voltage it will produce, the current limits, and other operating characteristics are built in. This information is usually stamped on a nameplate on the case so that the user will know the limitations.

Q6. How are alternators usually rated?

Q7. What type of prime mover requires a specially designed high-speed alternator?

Q8. Salient-pole rotors may be used in alternators driven by what types of prime movers?

SINGLE-PHASE ALTERNATORS

A generator that produces a single, continuously alternating voltage is known as a SINGLE-PHASE alternator. All of the alternators that have been discussed so far fit this definition. The stator (armature) windings are connected in series. The individual voltages, therefore, add to produce a single-phase ac voltage. Figure 3-5 shows a basic alternator with its single-phase output voltage.

The definition of phase as you learned it in studying ac circuits may not help too much right here. Remember, "out of phase" meant "out of time."

Now, it may be easier to think of the word phase as meaning voltage as in single voltage. The need for a modified definition of phase in this usage will be easier to see as we go along.

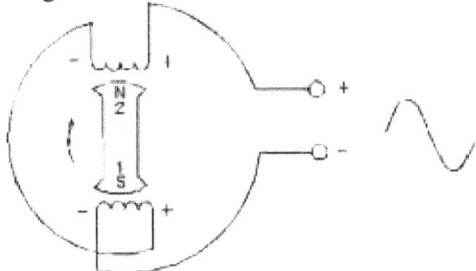

Figure 3-5.—Single-phase alternator.

Single-phase alternators are found in many applications. They are most often used when the loads being driven are relatively light. The reason for this will be more apparent as we get into multiphase alternators (also called polyphase).

Power that is used in homes, shops, and ships to operate portable tools and small appliances is single-phase power. Single-phase power alternators always generate single-phase power. However, all single-phase power does not come from single-phase alternators. This will sound more reasonable to you as we get into the next subjects.

Q9. What does the term single phase indicate?

Q10. In single-phase alternators, in order for the voltages induced in all the armature windings to add together for a single output, how must the windings be connected?

TWO-PHASE ALTERNATORS

Two phase implies two voltages if we apply our new definition of phase. And, it's that simple. A two-phase alternator is designed to produce two completely separate voltages. Each voltage, by itself, may be considered as a single-phase voltage. Each is generated completely independent of the other. Certain advantages are gained. These and the mechanics of generation will be covered in the following paragraphs.

Generation of Two-Phase Power

Figure 3-6 shows a simplified two-pole, two-phase alternator. Note that the windings of the two phases are physically at right angles (90°) to each other. You would expect the outputs of each phase to be 90° apart, which they are. The graph shows the two phases to be 90° apart, with A leading B. Note that by using our original definition of phase (from previous modules), we could say that A and B are 90° out of phase. There will always be 90° between the phases of a two-phase alternator. This is by design.

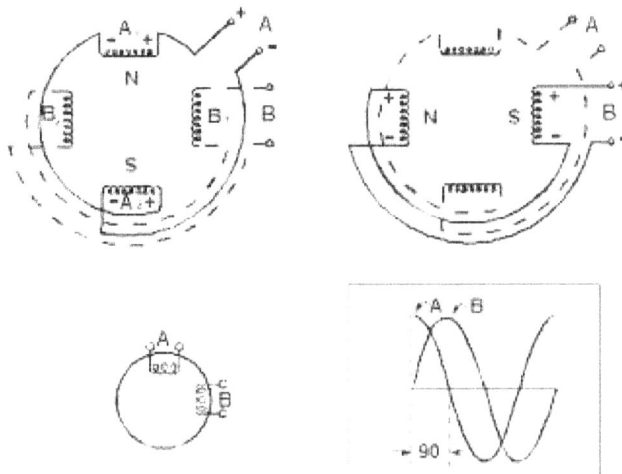

Figure 3-6.—Two-phase alternator.

Now, let's go back and see the similarities and differences between our original (single-phase) alternators and this new one (two-phase). Note that the principles applied are not new. This alternator works the same as the others we have discussed.

The stator in figure 3-6 consists of two single-phase windings completely separated from each other. Each winding is made up of two windings that are connected in series so that their voltages add. The rotor is identical to that used in the single-phase alternator. In the left-hand schematic, the rotor poles are opposite all the windings of phase A. Therefore, the voltage induced in phase A is maximum, and the voltage induced in phase B is zero. As the rotor continues rotating counterclockwise, it moves away from the A windings and approaches the B windings. As a result, the voltage induced in phase A decreases from its maximum value, and the voltage induced in phase B increases from zero. In the right-hand schematic, the rotor poles are opposite the windings of phase B. Now the voltage induced in phase B is maximum, whereas the voltage induced in phase A has dropped to zero. Notice that a 90-degree rotation of the rotor corresponds to one-quarter of a cycle, or 90 electrical degrees. The waveform picture shows the voltages induced in phase A and B for one cycle. The two voltages are 90° out of phase. Notice that the two outputs, A and B, are independent of each other. Each output is a single-phase voltage, just as if the other did not exist.

The obvious advantage, so far, is that we have two separate output voltages. There is some saving in having one set of bearings, one rotor, one housing, and so on, to do the work of two. There is the disadvantage of having twice as many stator coils, which require a larger and more complex stator.

The large schematic in figure 3-7 shows four separate wires brought out from the A and B stator windings. This is the same as in figure 3-6. Notice, however, that the dotted wire now connects one end of B1 to one end of A2. The effect of making this connection is to provide a new output voltage. This sine-wave voltage, C in the picture, is larger than either A or B. It is the result of adding the instantaneous values of phase A and phase B. For this reason it appears exactly half way between A and B. Therefore, C must lag A by 45° and lead B by 45°, as shown in the small vector diagram.

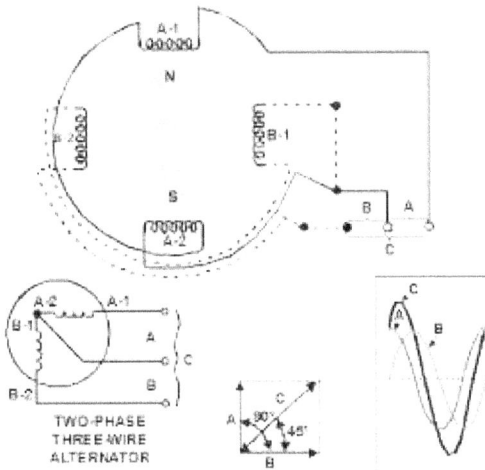

Figure 3-7.—Connections of a two-phase, three-wire alternator output.

Now, look at the smaller schematic diagram in figure 3-7. Only three connections have been brought out from the stator. Electrically, this is the same as the large diagram above it. Instead of being connected at the output terminals, the B1-A2 connection was made internally when the stator was wired. A two-phase alternator connected in this manner is called a two-phase, three-wire alternator.

The three-wire connection makes possible three different load connections: A and B (across each phase), and C (across both phases). The output at C is always 1.414 times the voltage of either phase. These multiple outputs are additional advantages of the two-phase alternator over the single-phase type.

Now, you can understand why single-phase power doesn't always come from single-phase alternators. It can be generated by two-phase alternators as well as other multiphase (polyphase) alternators, as you will soon see.

The two-phase alternator discussed in the preceding paragraphs is seldom seen in actual use. However, the operation of polyphase alternators is more easily explained using two phases than three phases. The three-phase alternator, which will be covered next, is by far the most common of all alternators in use today, both in military and civilian applications.

Q11. What determines the phase relationship between the voltages in a two-phase ac generator?

Q12. How many voltage outputs are available from a two-phase three-wire alternator?

Q13. What is the relationship of the voltage at C in figure 3-7 to the voltages at A and B?

THREE-PHASE ALTERNATOR

The three-phase alternator, as the name implies, has three single-phase windings spaced such that the voltage induced in any one phase is displaced by 120° from the other two. A schematic diagram of a three-phase stator showing all the coils becomes complex, and it is difficult to see what is actually happening. The simplified schematic of figure 3-8, view A, shows all the windings of each phase lumped together as one winding. The rotor is omitted for simplicity. The voltage waveforms generated across each phase are drawn on a graph, phase-displaced 120° from each other. The three-phase alternator as shown in this schematic is made up of three single-phase alternators whose generated voltages are out of phase by 120°. The three phases are independent of each other.

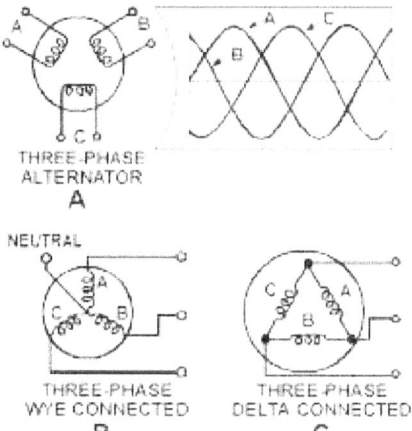

Figure 3-8.—Three-phase alternator connections.

Rather than having six leads coming out of the three-phase alternator, the same leads from each phase may be connected together to form a wye (Y) connection, as shown in figure 3-8, view B. It is called a wye connection because, without the neutral, the windings appear as the letter Y, in this case sideways or upside down.

The neutral connection is brought out to a terminal when a single-phase load must be supplied. Single-phase voltage is available from neutral to A, neutral to B, and neutral to C.

In a three-phase, Y-connected alternator, the total voltage, or line voltage, across any two of the three line leads is the vector sum of the individual phase voltages. Each line voltage is 1.73 times one of the phase voltages. Because the windings form only one path for current flow between phases, the line and phase currents are the same (equal).

A three-phase stator can also be connected so that the phases are connected end-to-end; it is now delta connected (fig. 3-8, view C). (Delta because it looks like the Greek letter delta, A.) In the delta connection, line voltages are equal to phase voltages, but each line current is equal to 1.73 times the phase current. Both the wye and the delta connections are used in alternators.

The majority of all alternators in use in the Navy today are three-phase machines. They are much more efficient than either two-phase or single-phase alternators.

Three-Phase Connections

The stator coils of three-phase alternators may be joined together in either wye or delta connections, as shown in figure 3-9. With these connections only three wires come out of the alternator. This allows convenient connection to three-phase motors or power distribution transformers. It is necessary to use three-phase transformers or their electrical equivalent with this type of system.

B   C   A
DELTA WYE CONNECTED CONNECTED

Figure 3-9.—Three-phase alternator or transformer connections.

A three-phase transformer may be made up of three, single-phase transformers connected in delta, wye, or a combination of both. If both the primary and secondary are connected in wye, the transformer is called a wye-wye. If both windings are connected in delta, the transformer is called a delta-delta.

Figure 3-10 shows single-phase transformers connected delta-delta for operation in a three-phase system. You will note that the transformer windings are not angled to illustrate the typical delta (Δ) as has been done with alternator windings. Physically, each transformer in the diagram stands alone. There is no angular relationship between the windings of the individual transformers. However, if you follow the connections, you will see that they form an electrical delta. The primary windings, for example, are connected to each other to form a closed loop. Each of these junctions is fed with a phase voltage from a three-phase alternator. The alternator may be connected either delta or wye depending on load and voltage requirements, and the design of the system.

THREE-PHASE OR THREE-PHASE SINGLE-PHASE INPUT (PRIMARY) OUTPUT (SECONDARY)

A o
B o
C o

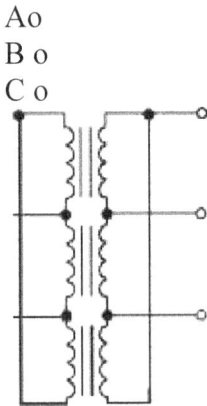

Figure 3-10.—Three single-phase transformers connected delta-delta.

Figure 3-11 shows three single-phase transformers connected wye-wye. Again, note that the transformer windings are not angled. Electrically, a Y is formed by the connections. The lower connections of each winding are shorted together. These form the common point of the wye. The opposite end of each winding is isolated. These ends form the arms of the wye.

THREE-PHASE INPUT (PRIMARY)

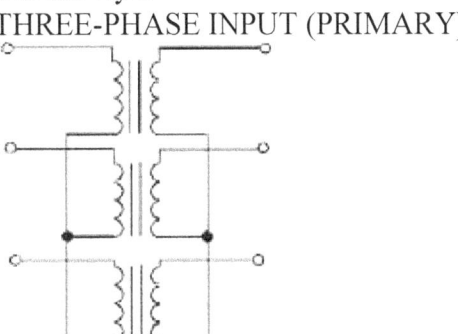

Figure 3-11.—Three single-phase transformers connected wye-wye.

The ac power on most ships is distributed by a three-phase, three-wire, 450-volt system. The single-phase transformers step the voltage down to 117 volts. These transformers are connected delta-delta as in figure 3-10. With a delta-delta configuration, the load may be a three-phase device connected to all phases; or, it may be a single-phase device connected to only one phase.

At this point, it is important to remember that such a distribution system includes everything between the alternator and the load. Because of the many choices that three-phase systems provide, care must be taken to ensure that any change of connections does not provide the load with the wrong voltage or the wrong phase.

Q14. In a three-phase alternator, what is the phase relationship between the individual output voltages?

Q15. What are the two methods of connecting the outputs from a three-phase alternator to the load?

Q16. Ships' generators produce 450-volt, three-phase, ac power; however, most equipment uses 117-volt, single-phase power What transformers and connections are used to convert 450-volt, three-phase power to 117-volt, single-phase power?

FREQUENCY

The output frequency of alternator voltage depends upon the speed of rotation of the rotor and the number of poles. The faster the speed, the higher the frequency. The lower the speed, the lower the frequency. The more poles there are on the rotor, the higher the frequency is for a given speed. When a rotor has rotated through an angle such that two adjacent rotor poles (a north and a south pole) have passed one winding, the voltage induced in that winding will have varied through one complete cycle. For a given frequency, the more pairs of poles there are, the lower the speed of rotation. This principle is

illustrated in figure 3-12; a two-pole generator must rotate at four times the speed of an eight-pole generator to produce the same frequency of generated voltage. The frequency of any ac generator in hertz (Hz), which is the number of cycles per second, is related to the number of poles and the speed of rotation, as expressed by the equation

T7 NP

r =

120

where P is the number of poles, N is the speed of rotation in revolutions per minute (rpm), and 120 is a constant to allow for the conversion of minutes to seconds and from poles to pairs of poles. For example, a 2-pole, 3600-rpm alternator has a frequency of 60 Hz; determined as follows:

2 x3600 ,,,,

= 60Hz

120

A 4-pole, 1800-rpm generator also has a frequency of 60 Hz. A 6-pole, 500-rpm generator has a frequency of

120

A 12-pole, 4000-rpm generator has a frequency of

12*4000

= 400Hz

120

Q17. What two factors determine the frequency of the output voltage of an alternator?

Q18. What is the frequency of the output voltage of an alternator with four poles that is rotated at 3600 rpm?

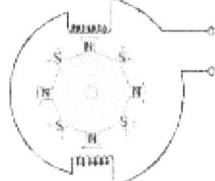

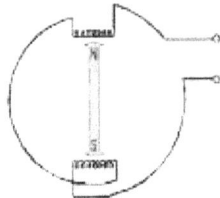

BOTH ALTERNATORS ARE ROTATING AT 120 RPM: F= ^

0° 180* 36C? 8-POLE LOW SPEED

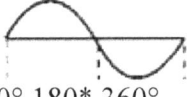

0° 180* 360°
2-POLE LOW SPEED

Figure 3-12.—Frequency regulation.

VOLTAGE REGULATION

As we have seen before, when the load on a generator is changed, the terminal voltage varies. The amount of variation depends on the design of the generator.

The voltage regulation of an alternator is the change of voltage from full load to no load, expressed as a percentage of full-load volts, when the speed and dc field current are held constant.

EfL "

100 = Percent of regulation

Assume the no-load voltage of an alternator is 250 volts and the full-load voltage is 220 volts. The percent of regulation is

250-220 220
x100 =13.6%

Remember, the lower the percent of regulation, the better it is in most applications.

Q19. The variation in output voltage as the load changes is referred to as what? How is it expressed?

PRINCIPLES OF AC VOLTAGE CONTROL

In an alternator, an alternating voltage is induced in the armature windings when magnetic fields of alternating polarity are passed across these windings. The amount of voltage induced in the windings

depends mainly on three things: (1) the number of conductors in series per winding, (2) the speed (alternator rpm) at which the magnetic field cuts the winding, and (3) the strength of the magnetic field. Any of these three factors could be used to control the amount of voltage induced in the alternator windings.

The number of windings, of course, is fixed when the alternator is manufactured. Also, if the output frequency is required to be of a constant value, then the speed of the rotating field must be held constant. This prevents the use of the alternator rpm as a means of controlling the voltage output. Thus, the only practical method for obtaining voltage control is to control the strength of the rotating magnetic field. The strength of this electromagnetic field may be varied by changing the amount of current flowing through the field coil. This is accomplished by varying the amount of voltage applied across the field cod.

Q20. How is output voltage controlled in practical alternators?

PARALLEL OPERATION OF ALTERNATORS

Alternators are connected in parallel to (1) increase the output capacity of a system beyond that of a single unit, (2) serve as additional reserve power for expected demands, or (3) permit shutting down one machine and cutting in a standby machine without interrupting power distribution. When alternators are of sufficient size, and are operating at different frequencies and terminal voltages, severe damage may result if they are suddenly connected to each other through a common bus. To avoid this, the machines must be synchronized as closely as possible before connecting them together. This may be accomplished by connecting one generator to the bus (referred to as bus generator), and then synchronizing the other (incoming generator) to it before closing the incoming generator's main power contactor. The generators are synchronized when the following conditions are set:

1. Equal terminal voltages. This is obtained by adjustment of the incoming generator's field strength.

2. Equal frequency. This is obtained by adjustment of the incoming generator's prime-mover speed.

3. Phase voltages in proper phase relation. The procedure for synchronizing generators is not discussed in this chapter. At this point, it is enough for you to know that the above must be accomplished to prevent damage to the machines.

Q21. What generator characteristics must be considered when alternators are synchronized for parallel operation?

SUMMARY

This chapter has presented an introduction to the subject of alternators. You have studied the characteristics and applications of different types. The following information provides a summary of the chapter for your review.

MAGNETIC INDUCTION is the process of inducing an emf in a coil whenever the coil is placed in a magnetic field and motion exists between the coil and the magnetic lines of flux. This is true if either the coil or the magnetic field moves, as long as the coil is caused to cut across magnetic flux lines.

The ROTATING ARMATURE-ALTERNATOR is essentially a loop rotating through a stationary magnetic fealties cutting action of the loop through the magnetic field generates ac in the loop. This ac is removed from the loop by means of slip rings and applied to an external load.

Ill
FIELD EXCITATION!
FIELD

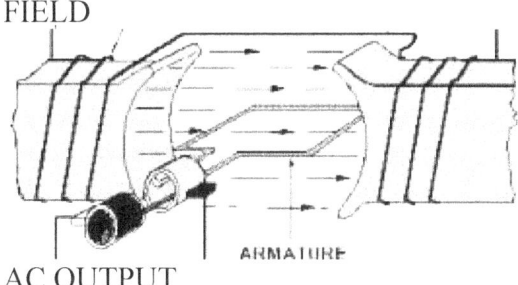

AC OUTPUT

The ROTATING-FIELD ALTERNATOR has a stationary armature and a rotating field. High voltages can be generated in the armature and applied to the load directly, without the need of slip rings and brushes. The low dc voltage is applied to the rotor field by means of slip rings, but this does not introduce any insulation problems.

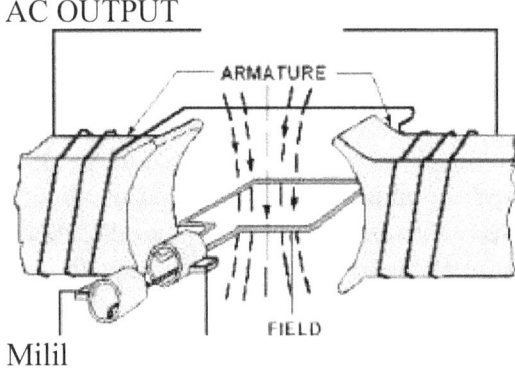

Milil

ROTOR CONSTRUCTION in alternators may be either of two types. The salient-pole rotor is used in slower speed alternators. The turbine driven-type is wound in a manner to allow high-speed use without flying apart.

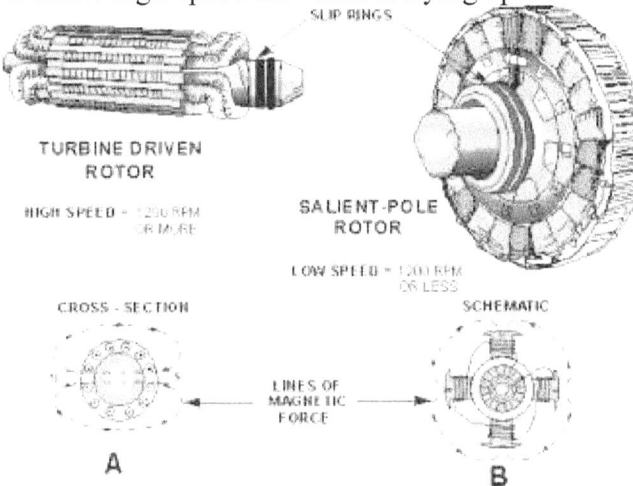

GENERATOR RATINGS are dependent on the amount of current they are capable of providing at full output voltage; this rating is expressed as the product of the voltage times the current. A 10-volt alternator capable of supplying 10 amperes of current would be rated at 100 volt-amperes. Larger alternators are rated in kilovolt-amperes.

EXCITER GENERATORS are small dc generators built into alternators to provide excitation current to field windings. These dc generators are called exciters.

The SINGLE-PHASE ALTERNATOR has an armature that consists of a number of windings placed symmetrically around the stator and connected in series. The voltages generated in each winding add to produce the total voltage across the two output terminals.

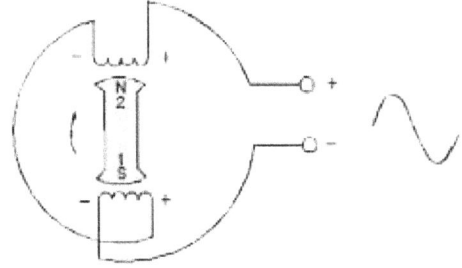

A TWO-PHASE ALTERNATOR consists of two phases whose windings are so placed around the stator that the voltages generated in them are 90° out of phase.

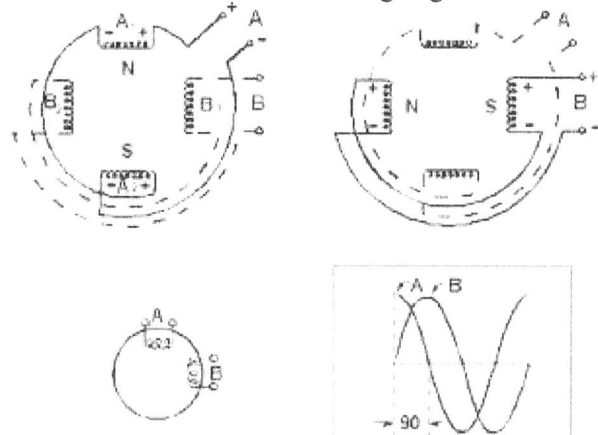

TWO-PHASE ALTERNATOR CONNECTIONS may be modified so that the output of a two-phase alternator is in a three-wire manner, which actually provides three outputs, two induced phase voltages, plus a vectorial sum voltage.

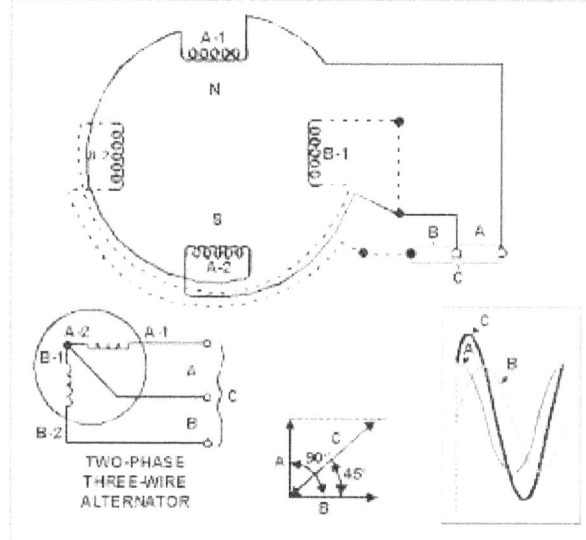

In THREE-PHASE ALTERNATORS the windings have voltages generated in them which are 120° out of phase. Three-phase alternators are most often used to generate ac power.

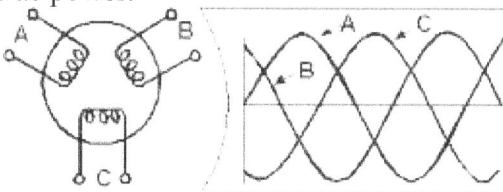

C a
THREE-PHASE ALTERNATOR

THREE-PHASE ALTERNATOR CONNECTIONS may be delta or wye connections depending on the application. The ac power aboard ship is usually taken from the ship's generators through delta connections, for the convenience of step-down transformers.

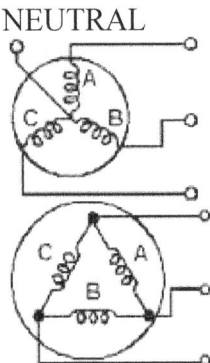

THREE-PHASE THREE-PHASE WYE CONNECTED DELTA CONNECTED

ALTERNATOR FREQUENCY depends upon the speed of rotation and the number of pairs of rotor poles.

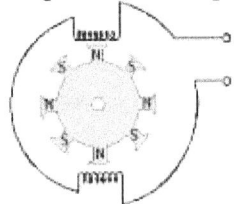

BOTH ALTERNATORS ARE ROTATING AT 120 RPM: $F = \dfrac{NP}{120}$

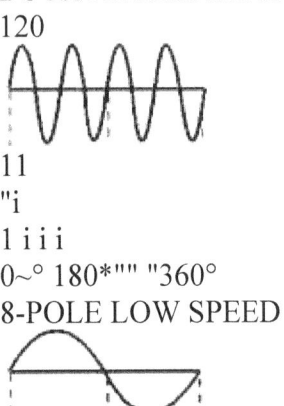

11
"i
1 i i i
0~° 180*"" "360°
8-POLE LOW SPEED

0*" 180* "360* 2-POLE LOW SPEED

VOLTAGE REGULATION is the change in output voltage of an alternator under varying load conditions.

VOLTAGE CONTROL in alternators is accomplished by varying the current in the field windings, much as in dc generators.

ANSWERS TO QUESTIONS Ql. THROUGH Q21.

A1. A conductor and a magnetic field.

A2. Armature.

A3. Rotating armature and rotating field.

A4. Output voltage is taken directly from the armature (not through brushes or slip rings).

A5. To provide dc current for the rotating field.

A6. Kilovolt-amperes (volt amperes).

A7. Steam turbine.

A8. Internal combustion engines, water force and electric motors.

A9. One voltage (one output).

A10. In series.
A11. Placement of armature coils.
A12. Three.
A13. C is 1.414 times greater than A or B.
A14. Each phase is displaced 120° from the other two.
A15. Wye and Delta.
A16. Three single-phase, delta-delta, step-down transformers.
A17. Speed of rotation and number of poles.
A18. 120 Hz.
A19. Voltage regulation. As a percentage.
A20. By varying the voltage applied to the field windings.
A21. Output voltage, frequency, and phase relationships.

CHAPTER 4
ALTERNATING CURRENT MOTORS

LEARNING OBJECTIVES

Upon completion of this chapter you will be able to:

1. List three basic types of ac motors and describe the characteristics of each type.
2. Describe the characteristics of a series motor that enable it to be used as a universal motor.
3. Explain the relationships of the individual phases of multiphase voltages as they produce rotating magnetic fields in ac motors.
4. Describe the placement of stator windings in two-phase, ac motors using rotating fields.
5. List the similarities and differences between the stator windings of two-phase and three-phase ac motors.
6. State the primary application of synchronous motors, and explain the characteristics that make them suitable for that application.
7. Describe the features that make the ac induction motor the most widely used of electric motors.
8. Describe the difference between the rotating field of multiphase motors and the "apparent" rotating field of single-phase motors.
9. Explain the operation of split-phase windings in single-phase ac induction motors. 10. Describe the effects of shaded poles in single-phase, ac induction motors.

INTRODUCTION

Most of the power-generating systems, ashore and afloat, produce ac. For this reason a majority of the motors used throughout the Navy are designed to operate on ac. There are other advantages in the use of ac motors besides the wide availability of ac power. In general, ac motors cost less than dc motors. Some types of ac motors do not use brushes and commutators. This eliminates many problems of maintenance and wear. It also eliminates the problem of dangerous sparking.

An ac motor is particularly well suited for constant-speed applications. This is because its speed is determined by the frequency of the ac voltage applied to the motor terminals.

The dc motor is better suited than an ac motor for some uses, such as those that require variable-speeds. An ac motor can also be made with variable speed characteristics but only within certain limits.

Industry builds ac motors in different sizes, shapes, and ratings for many different types of jobs. These motors are designed for use with either polyphase or single-phase power systems. It is not possible here to cover all aspects of the subject of ac motors. Only the principles of the most commonly used types are dealt with in this chapter.

In this chapter, ac motors will be divided into (1) series, (2) synchronous, and (3) induction motors. Single-phase and polyphase motors will be discussed. Synchronous motors, for purposes of this chapter, may be considered as polyphase motors, of constant speed, whose rotors are energized with dc voltage. Induction motors, single-phase or polyphase, whose rotors are energized by induction, are the most commonly used ac motor. The series ac motor, in a sense, is a familiar type of motor. It is very similar to the dc motor that was covered in chapter 2 and will serve as a bridge between the old and the new.

Q1. What are the three basic types of ac motors?

SERIES AC MOTOR

A series ac motor is the same electrically as a dc series motor. Refer to figure 4-1 and use the left-hand rule for the polarity of coils. You can see that the instantaneous magnetic polarities of the armature and field oppose each other, and motor action results. Now, reverse the current by reversing the polarity of the input. Note that the field magnetic polarity still opposes the armature magnetic polarity. This is because the reversal effects both the armature and the field. The ac input causes these reversals to take place continuously.

FIELD COIL

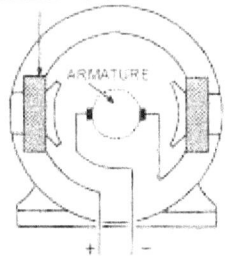

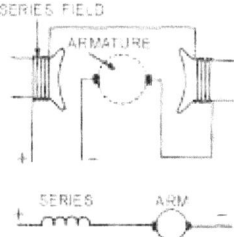

Figure 4-1.—Series ac motor.

The construction of the ac series motor differs slightly from the dc series motor. Special metals, laminations, and windings are used. They reduce losses caused by eddy currents, hysteresis, and high reactance. Dc power can be used to drive an ac series motor efficiently, but the opposite is not true.

The characteristics of a series ac motor are similar to those of a series dc motor. It is a varying-speed machine. It has low speeds for large loads and high speeds for light loads. The starting torque is very

high. Series motors are used for driving fans, electric drills, and other small appliances. Since the series ac motor has the same general characteristics as the series dc motor, a series motor has been designed that can operate both on ac and dc. This ac/dc motor is called a universal motor. It finds wide use in small electric appliances. Universal motors operate at lower efficiency than either the ac or dc series motor. They are built in small sizes only. Universal motors do not operate on polyphase ac power.

Q2. Series motors are generally used to operate what type of equipment?

Q3. Why are series motors sometimes called universal motors?

ROTATING MAGNETIC FIELDS

The principle of rotating magnetic fields is the key to the operation of most ac motors. Both synchronous and induction types of motors rely on rotating magnetic fields in their stators to cause their rotors to turn.

The idea is simple. A magnetic field in a stator can be made to rotate electrically, around and around. Another magnetic field in the rotor can be made to chase it by being attracted and repelled by the stator field. Because the rotor is free to turn, it follows the rotating magnetic field in the stator. Let's see how it is done.

Rotating magnetic fields may be set up in two-phase or three-phase machines. To establish a rotating magnetic field in a motor stator, the number of pole pairs must be the same as (or a multiple of) the number of phases in the applied voltage. The poles must then be displaced from each other by an angle equal to the phase angle between the individual phases of the applied voltage.

Q4. What determines the number of field poles required to establish a rotating magnetic field in a multiphase motor stator?

TWO-PHASE ROTATING MAGNETIC FIELD

A rotating magnetic field is probably most easily seen in a two-phase stator. The stator of a two-phase induction motor is made up of two windings (or a multiple of two). They are placed at right angles to each other around the stator. The simplified drawing in figure 4-2 illustrates a two-phase stator.

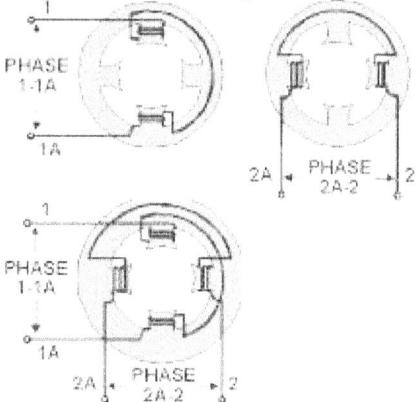

Figure 4-2.—Two-phase motor stator.

If the voltages applied to phases 1-1A and 2-2A are 90° out of phase, the currents that flow in the phases are displaced from each other by 90°. Since the magnetic fields generated in the coils are in phase with their respective currents, the magnetic fields are also 90° out of phase with each other. These two out-of-phase magnetic fields, whose coil axes are at right angles to each other, add together at every instant during their cycle. They produce a resultant field that rotates one revolution for each cycle of ac.

To analyze the rotating magnetic field in a two-phase stator, refer to figure 4-3.

The arrow represents the rotor. For each point set up on the voltage chart, consider that current flows in a direction that will cause the magnetic polarity indicated at each pole piece. Note that from one point to the next, the polarities are rotating from one pole to the next in a clockwise manner. One complete cycle of input voltage produces a 360-degree rotation of the pole polarities. Let's see how this result is obtained.

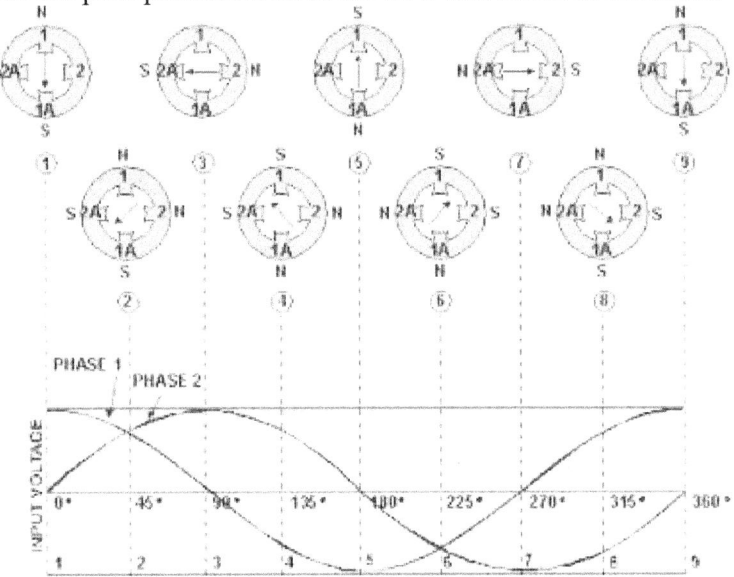

Figure 4-3.—Two-phase rotating field.

The waveforms in figure 4-3 are of the two input phases, displaced 90° because of the way they were generated in a two-phase alternator. The waveforms are numbered to match their associated phase. Although not shown in this figure, the windings for the poles 1-1A and 2-2A would be as shown in the previous figure. At position 1, the current flow and magnetic field in winding 1-1A is at maximum (because the phase voltage is maximum). The current flow and magnetic field in winding 2-2A is zero (because the phase voltage is zero). The resultant magnetic field is therefore in the direction of the 1-1A axis. At the 45-degree point (position 2), the resultant magnetic field lies midway between windings 1-1A and 2-2A. The coil currents and magnetic fields are equal in strength. At 90° (position 3), the magnetic field in winding 1-1A is zero. The magnetic field in winding 2-2A is at maximum. Now the resultant magnetic field lies along the axis of the 2-2A winding as shown. The resultant magnetic field has rotated clockwise through 90° to get from position 1 to position 3. When the two-phase voltages have completed one full cycle (position 9), the resultant magnetic field has rotated through 360°. Thus, by placing two windings at right angles to each other and exciting these windings with voltages 90° out of phase, a rotating magnetic field results.

Two-phase motors are rarely used except in special-purpose equipment. They are discussed here to aid in understanding rotating fields. You will, however, encounter many single-phase and three-phase motors.

Q5. What is the angular displacement between field poles in a two-phase motor stator?

THREE-PHASE ROTATING FIELDS

The three-phase induction motor also operates on the principle of a rotating magnetic field. The following discussion shows how the stator windings can be connected to a three-phase ac input and have a resultant magnetic field that rotates.

Figure 4-4, views A-C show the individual windings for each phase. Figure 4-4, view D, shows how the three phases are tied together in a Y-connected stator. The dot in each diagram indicates the common point of the Y-connection. You can see that the individual phase windings are equally spaced around the stator. This places the windings 120° apart.

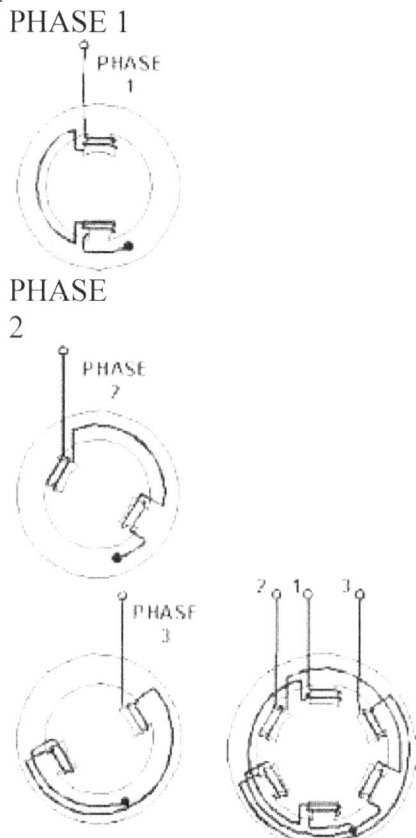

Figure 4-4.—Three-phase, Y-connected stator.

The three-phase input voltage to the stator of figure 4-4 is shown in the graph of figure 4-5. Use the left-hand rule for determining the electromagnetic polarity of the poles at any given instant. In applying the rule to the coils in figure 4-4, consider that current flows toward the terminal numbers for positive voltages, and away from the terminal numbers for negative voltages.

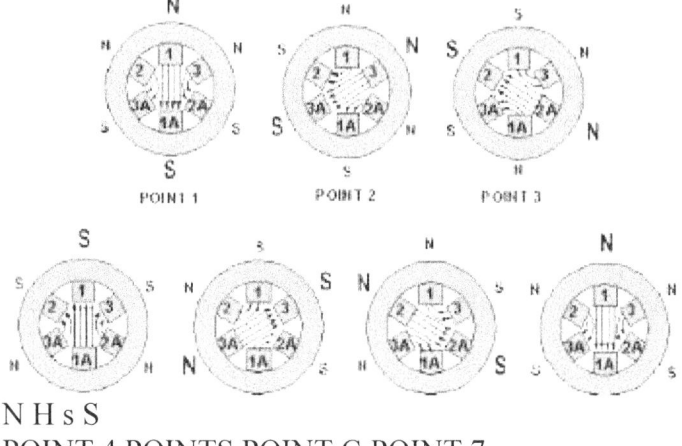

Figure 4-5.—Three-phase rotating-field polarities and input voltages.

The results of this analysis are shown for voltage points 1 through 7 in figure 4-5. At point 1, the magnetic field in coils 1-1A is maximum with polarities as shown. At the same time, negative voltages are being felt in the 2-2A and 3-3A windings. These create weaker magnetic fields, which tend to aid the 1-1A field. At point 2, maximum negative voltage is being felt in the 3-3A windings. This creates a strong magnetic field which, in turn, is aided by the weaker fields in 1-1A and 2-2A. As each point on the voltage graph is analyzed, it can be seen that the resultant magnetic field is rotating in a clockwise direction. When the three-phase voltage completes one full cycle (point 7), the magnetic field has rotated through 360°.

Q6. What is the major difference between a two-phase and a three-phase stator?

ROTOR BEHAVIOR IN A ROTATING FIELD

For purposes of explaining rotor movement, let's assume that we can place a bar magnet in the center of the stator diagrams of figure 4-5. We'll mount this magnet so that it is free to rotate in this area. Let's also assume that the bar magnet is aligned so that at point 1 its south pole is opposite the large N of the stator field.

You can see that this alignment is natural. Unlike poles attract, and the two fields are aligned so that they are attracting. Now, go from point 1 through point 7. As before, the stator field rotates clockwise. The bar magnet, free to move, will follow the stator field, because the attraction between the two fields

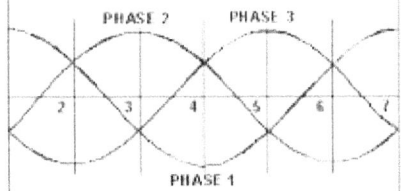

continues to exist. A shaft running through the pivot point of the bar magnet would rotate at the same speed as the rotating field. This speed is known as synchronous speed. The shaft represents the shaft of an operating motor to which the load is attached.

Remember, this explanation is an oversimplification. It is meant to show how a rotating field can cause mechanical rotation of a shaft. Such an arrangement would work, but it is not used. There are limitations to a permanent magnet rotor. Practical motors use other methods, as we shall see in the next paragraphs.

SYNCHRONOUS MOTORS

The construction of the synchronous motors is essentially the same as the construction of the salient-pole alternator. In fact, such an alternator may be run as an ac motor. It is similar to the drawing in figure 4-6. Synchronous motors have the characteristic of constant speed between no load and full load. They are capable of correcting the low power factor of an inductive load when they are operated under certain conditions. They are often used to drive dc generators. Synchronous motors are designed in sizes up to thousands of horsepower. They may be designed as either single-phase or multiphase machines. The discussion that follows is based on a three-phase design.

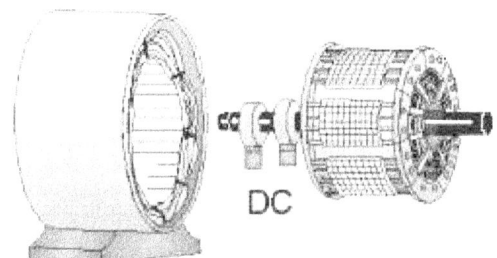

Figure 4-6.—Revolving-field synchronous motor.

To understand how the synchronous motor works, assume that the application of three-phase ac power to the stator causes a rotating magnetic field to be set up around the rotor. The rotor is energized with dc (it acts like a bar magnet). The strong rotating magnetic field attracts the strong rotor field activated by the dc. This results in a strong turning force on the rotor shaft. The rotor is therefore able to turn a load as it rotates in step with the rotating magnetic field.

It works this way once it's started. However, one of the disadvantages of a synchronous motor is that it cannot be started from a standstill by applying three-phase ac power to the stator. When ac is applied to the stator, a high-speed rotating magnetic field appears immediately. This rotating field rushes past the rotor poles so quickly that the rotor does not have a chance to get started. In effect, the rotor is repelled first in one direction and then the other. A synchronous motor in its purest form has no starting torque. It has torque only when it is running at synchronous speed.

A squirrel-cage type of winding is added to the rotor of a synchronous motor to cause it to start. The squirrel cage is shown as the outer part of the rotor in figure 4-7. It is so named because it is shaped and looks something like a turnable squirrel cage. Simply, the windings are heavy copper bars shorted together by copper rings. A low voltage is induced in these shorted windings by the rotating three-phase stator field. Because of the short circuit, a relatively large current flows in the squirrel cage. This causes a magnetic field that interacts with the rotating field of the stator. Because of the interaction, the rotor begins to turn, following the stator field; the motor starts. We will run into squirrel cages again in other applications, where they will be covered in more detail.

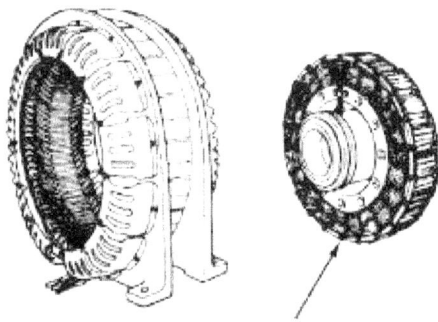

SQUIRREL-CAGE WINDING OVER SALIENT-POLE WINDINGS
Figure 4-7.—Self-starting synchronous ac motor.

To start a practical synchronous motor, the stator is energized, but the dc supply to the rotor field is not energized. The squirrel-cage windings bring the rotor to near synchronous speed. At that point, the dc field is energized. This locks the rotor in step with the rotating stator field. Full torque is developed, and the load is driven. A mechanical switching device that operates on centrifugal force is often used to apply dc to the rotor as synchronous speed is reached.

The practical synchronous motor has the disadvantage of requiring a dc exciter voltage for the rotor. This voltage may be obtained either externally or internally, depending on the design of the motor.

Q7. What requirement is the synchronous motor specifically designed to meet?

INDUCTION MOTORS

The induction motor is the most commonly used type of ac motor. Its simple, rugged construction costs relatively little to manufacture. The induction motor has a rotor that is not connected to an external source of voltage. The induction motor derives its name from the fact that ac voltages are induced in the rotor circuit by the rotating magnetic field of the stator. In many ways, induction in this motor is similar to the induction between the primary and secondary windings of a transformer.

Large motors and permanently mounted motors that drive loads at fairly constant speed are often induction motors. Examples are found in washing machines, refrigerator compressors, bench grinders, and table saws.

The stator construction of the three-phase induction motor and the three-phase synchronous motor are almost identical. However, their rotors are completely different (see fig. 4-8). The induction rotor is made of a laminated cylinder with slots in its surface. The windings in these slots are one of two types (shown in fig. 4-9). The most common is the squirrel-cage winding. This entire winding is made up of

heavy copper bars connected together at each end by a metal ring made of copper or brass. No insulation is required between the core and the bars. This is because of the very low voltages generated in the rotor bars. The other type of winding contains actual coils placed in the rotor slots. The rotor is then called a wound rotor.

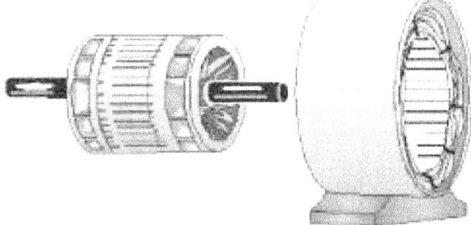

Figure 4-8.—Induction motor.

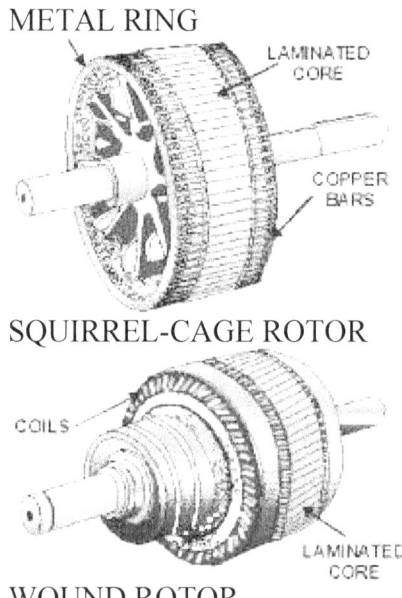

METAL RING
SQUIRREL-CAGE ROTOR
WOUND ROTOR

Figure 4-9.—Types of ac induction motor rotors.

Regardless of the type of rotor used, the basic principle is the same. The rotating magnetic field generated in the stator induces a magnetic field in the rotor. The two fields interact and cause the rotor to

turn. To obtain maximum interaction between the fields, the air gap between the rotor and stator is very small.

As you know from Lenz's law, any induced emf tries to oppose the changing field that induces it. In the case of an induction motor, the changing field is the motion of the resultant stator field. A force is exerted on the rotor by the induced emf and the resultant magnetic field. This force tends to cancel the relative motion between the rotor and the stator field. The rotor, as a result, moves in the same direction as the rotating stator field.

It is impossible for the rotor of an induction motor to turn at the same speed as the rotating magnetic field. If the speeds were the same, there would be no relative motion between the stator and rotor fields; without relative motion there would be no induced voltage in the rotor. In order for relative motion to exist between the two, the rotor must rotate at a speed slower than that of the rotating magnetic field. The. difference between the speed of the rotating stator field and the rotor speed is called slip. The smaller the slip, the closer the rotor speed approaches the stator field speed.

The speed of the rotor depends upon the torque requirements of the load. The bigger the load, the stronger the turning force needed to rotate the rotor. The turning force can increase only if the rotor-induced emf increases. This emf can increase only if the magnetic field cuts through the rotor at a faster rate. To increase the relative speed between the field and rotor, the rotor must slow down. Therefore, for heavier loads the induction motor turns slower than for lighter loads. You can see from the previous statement that slip is directly proportional to the load on the motor. Actually only a slight change in speed is necessary to produce the usual current changes required for normal changes in load. This is because the rotor windings have such a low resistance. As a result, induction motors are called constant-speed motors.

Q8. Why is the ac induction motor used more often than other types?

Q9. The speed of the rotor is always somewhat less than the speed of the rotating field. What is the difference called?

Q10. What determines the amount of slip in an induction motor?

SINGLE-PHASE INDUCTION MOTORS

There are probably more single-phase ac induction motors in use today than the total of all the other types put together.

It is logical that the least expensive, lowest maintenance type of ac motor should be used most often. The single-phase ac induction motor fits that description.

Unlike polyphase induction motors, the stator field in the single-phase motor does not rotate. Instead it simply alternates polarity between poles as the ac voltage changes polarity.

Voltage is induced in the rotor as a result of magnetic induction, and a magnetic field is produced around the rotor. This field will always be in opposition to the stator field (Lenz's law applies). The interaction between the rotor and stator fields will not produce rotation, however. The interaction is shown by the double-ended arrow in figure 4-10, view A. Because this force is across the rotor and through the pole pieces, there is no rotary motion, just a push and/or pull along this line.

N_R, S_R = rotor field
N_S, S_S = STATOR FIELD

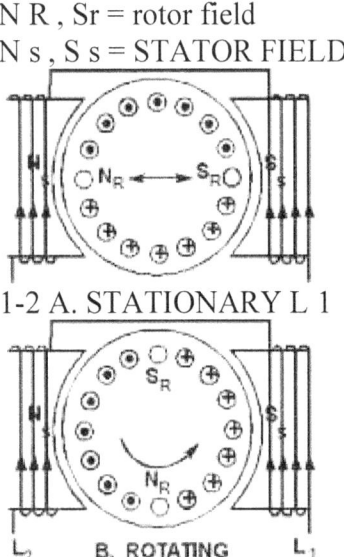

Figure 4-10.—Rotor currents in a single-phase ac induction motor.

Now, if the rotor is rotated by some outside force (a twist of your hand, or something), the push-pull along the line in figure 4-10, view A, is disturbed. Look at the fields as shown in figure 4-10, view B. At this instant the south pole on the rotor is being attracted by the left-hand pole. The north rotor pole is being attracted to the right-hand pole. All of this is a result of the rotor being rotated 90° by the outside force. The pull that now exists between the two fields becomes a rotary force, turning the rotor toward magnetic correspondence with the stator. Because the two fields continuously alternate, they will never actually line up, and the rotor will continue to turn once started. It remains for us to learn practical methods of getting the rotor to start.

There are several types of single-phase induction motors in use today. Basically they are identical except for the means of starting. In this chapter we will discuss the split-phase and shaded-pole motors; so named because of the methods employed to get them started. Once they are up to operating speed, all single-phase induction motors operate the same.

Q11. What type of ac motor is most widely used?

Split-Phase Induction Motors

One type of induction motor, which incorporates a starting device, is called a split-phase induction motor. Split-phase motors are designed to use inductance, capacitance, or resistance to develop a starting torque. The principles are those that you learned in your study of alternating current.

CAPACITOR-START.—The first type of split-phase induction motor that will be covered is the capacitor-start type. Figure 4-11 shows a simplified schematic of a typical capacitor-start motor. The stator consists of the main winding and a starting winding (auxiliary). The starting winding is connected in parallel with the main winding and is placed physically at right angles to it. A 90-degree electrical

phase difference between the two windings is obtained by connecting the auxiliary winding in series with a capacitor and starting switch. When the motor is first energized, the starting switch is closed. This places the capacitor in series with the auxiliary winding. The capacitor is of such value that the auxiliary circuit is effectively a resistive-capacitive circuit (referred to as capacitive reactance and expressed as X_c). In this circuit the current leads the line voltage by about 45° (because X_c about equals R). The main winding has enough resistance-inductance (referred to as inductive reactance and expressed as X_L) to cause the current to lag the line voltage by about 45° (because X_L about equals R). The currents in each winding are therefore 90° out of phase - so are the magnetic fields that are generated. The effect is that the two windings act like a two-phase stator and produce the rotating field required to start the motor.

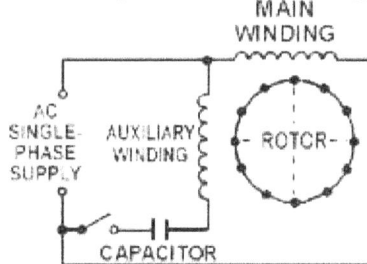

Figure 4-11.—Capacitor-start, ac induction motor.

When nearly full speed is obtained, a centrifugal device (the starting switch) cuts out the starting winding. The motor then runs as a plain single-phase induction motor. Since the auxiliary winding is only a light winding, the motor does not develop sufficient torque to start heavy loads. Split-phase motors, therefore, come only in small sizes.

RESISTANCE-START.—Another type of split-phase induction motor is the resistance-start motor. This motor also has a starting winding (shown in fig. 4-12) in addition to the main winding. It is switched in and out of the circuit just as it was in the capacitor-start motor. The starting winding is positioned at right angles to the main winding. The electrical phase shift between the currents in the two windings is obtained by making the impedance of the windings unequal. The main winding has a high inductance and a low resistance. The current, therefore, lags the voltage by a large angle. The starting winding is designed to have a fairly low inductance and a high resistance. Here the current lags the voltage by a smaller angle. For example, suppose the current in the main winding lags the voltage by 70°. The current in the auxiliary winding lags the voltage by 40°. The currents are, therefore, out of phase by 30°. The magnetic fields are out of phase by the same amount. Although the ideal angular phase difference is 90° for maximum starting torque, the 30-degree phase difference still generates a rotating field. This supplies enough torque to start the motor. When the motor comes up to speed, a speed-controlled switch disconnects the starting winding from the line, and the motor continues to run as an induction motor. The starting torque is not as great as it is in the capacitor-start.

MAIN WINDING
AC SINGLE-PHASE SUPPLY

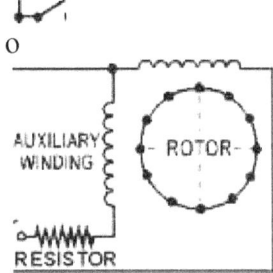

Figure 4-12.—Resistance-start ac induction motor.

Q12. How do split-phase induction motors become self-starting?

Shaded-Pole Induction Motors

The shaded-pole induction motor is another single-phase motor. It uses a unique method to start the rotor turning. The effect of a moving magnetic field is produced by constructing the stator in a special way. This motor has projecting pole pieces just like some dc motors. In addition, portions of the pole piece surfaces are surrounded by a copper strap called a shading coil. A pole piece with the strap in place is shown in figure 4-13. The strap causes the field to move back and forth across the face of the pole piece. Note the numbered sequence and points on the magnetization curve in the figure. As the alternating stator field starts increasing from zero (1), the lines of force expand across the face of the pole piece and cut through the strap. A voltage is induced in the strap. The current that results generates a field that opposes the cutting action (and decreases the strength) of the main field. This produces the following actions: As the field increases from zero to a maximum at 90°, a large portion of the magnetic lines of force are concentrated in the unshaded portion of the pole (1). At 90° the field reaches its maximum value. Since the lines of force have stopped expanding, no emf is induced in the strap, and no opposing magnetic field is generated. As a result, the main field is uniformly distributed across the pole (2). From 90° to 180°, the main field starts decreasing or collapsing inward. The field generated in the strap opposes the collapsing field. The effect is to concentrate the lines of force in the shaded portion of the pole face (3). You can see that from 0° to 180°, the main field has shifted across the pole face from the unshaded to the shaded portion. From 180° to 360°, the main field goes through the same change as it did from 0° to 180°; however, it is now in the opposite direction (4). The direction of the field does not affect the way the shaded pole works. The motion of the field is the same during the second half-cycle as it was during the first half of the cycle.

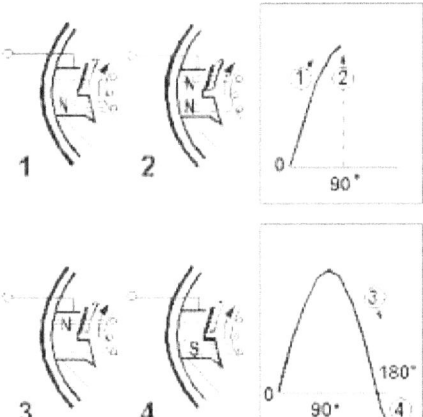

Figure 4-13.—Shaded poles as used in shaded-pole ac induction motors.

The motion of the field back and forth between shaded and unshaded portions produces a weak torque to start the motor. Because of the weak starting torque, shaded-pole motors are built only in small sizes. They drive such devices as fans, clocks, blowers, and electric razors.

Q13. Why are shaded-pole motors used to drive only very small devices?

Speed of Single-Phase Induction Motors

The speed of induction motors is dependent on motor design. The synchronous speed (the speed at which the stator field rotates) is determined by the frequency of the input ac power and the number of poles in the stator. The greater the number of poles, the slower the synchronous speed. The higher the frequency of applied voltage, the higher the synchronous speed. Remember, however, that neither frequency nor number of poles are variables. They are both fixed by the manufacturer.

The relationship between poles, frequency, and synchronous speed is as follows:

$$n \text{ (rpm)} = \frac{120f}{P}$$

where n is the synchronous speed in rpm, f is the frequency of applied voltage in hertz, and p is the number of poles in the stator.

Let's use an example of a 4-pole motor, built to operate on 60 hertz. The synchronous speed is determined as follows:

$$n = \frac{120f}{P}$$

$$n = \frac{120 \times 60}{4}$$

$$n = 1800 \text{ rpm}$$

Common synchronous speeds for 60-hertz motors are 3600, 1800, 1200, and 900 rpm, depending on the number of poles in the original design.

As we have seen before, the rotor is never able to reach synchronous speed. If it did, there would be no voltage induced in the rotor. No torque would be developed. The motor would not operate. The difference between rotor speed and synchronous speed is called slip. The difference between these two speeds is not great. For example, a rotor speed of 3400 to 3500 rpm can be expected from a synchronous speed of 3600 rpm.

SUMMARY

This chapter introduced you to the basic principles concerning ac motors. While many variations of types exist, the three types presented provide you with background for further study if you require more extensive knowledge of the subject. The following information provides a summary of the major subjects of this chapter for your review.

The three AC MOTOR TYPES presented are the series, synchronous, and induction ac motors.

AC SERIES MOTORS are nearly identical to the dc series motors. Special construction techniques allow ac series motors to be used as UNIVERSAL MOTORS, operating on either ac or dc power.

FIELD COIL

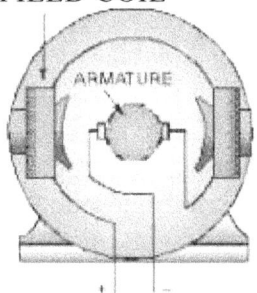

ROTATING FIELDS are developed by applying multiphase voltages to stator windings, which consist of multiple field coils. This rotating magnetic field causes the rotor to be pushed and pulled because of interaction between it and the rotor's own field.

TWO-PHASE ROTATING FIELDS require two pairs of field coils displaced by 90°. They must be energized by voltages that also have a phase displacement of 90°.

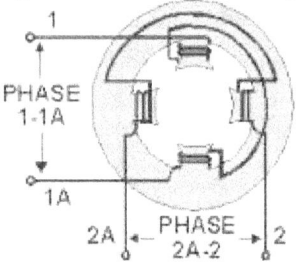

THREE-PHASE ROTATING FIELDS require three pairs of windings 120° apart, energized by voltages that also have a 120-degree phase displacement.

SYNCHRONOUS MOTORS are specifically designed to maintain constant speed, with the rotor synchronous to the rotating field. Synchronous motors require modification (such as squirrel-cage windings) to be self-starting.

I.1FTAI n v.

SQUIRREL-CAGE ROTOR

WOUND ROTOR

INDUCTION MOTORS are the most commonly used of all electric motors due to their simplicity and low cost. Induction motors may be single-phase or multiphase. They do not require electrical rotor connection. Split-phase motors with special starting windings, and shaded-pole motors, are types of single-phase induction motors.

SYNCHRONOUS SPEED is the speed of stator field rotation. It is determined by the number of poles and the frequency of the input voltage. Thus, for a given motor, synchronous speed is constant.

SLIP is the difference between actual rotor speed and the synchronous speed in induction motors. Slip must exist for there to be torque at the rotor shaft.

ANSWERS TO QUESTIONS Ql. THROUGH Q13.

A1. Series, synchronous, induction.

A2. To power small appliances.

A3. They operate on either ac or dc.

A4. The number of phases in the applied voltage.

A5. 90°.

A6. Number and location of field poles.

A7. Constant speed required by some loads.

A8. They are simple and inexpensive to make.

A9. Slip.

A10. Load.

All. Single-phase induction motor.

All. By using combinations of inductance and capacitance to apply out-of phase currents in starting windings.

A13. They have very weak starting torques.

APPENDIX I

GLOSSARY

AMPLIDYNE—A special dc generator in which a small dc voltage applied to field windings controls a large output voltage from the generator. In effect, an amplidyne is a rotary amplifier, oftentimes producing gain in the order of 10,000.

ARMATURE—The windings in which the output voltage is generated in a generator or in which input current creates a magnetic field that interacts with the main field in a motor. Note: Armature is often used as being identical with ROTOR. This usage is correct only part of the time. See the text and the entries under ROTOR and STATOR in this Glossary.

ARMATURE LOSSES—Copper losses, eddy current losses, hysteresis losses which act to decrease the efficiency of armatures.

ARMATURE REACTION—The effect, in a dc generator, of current in the annature creating a magnetic field that distorts the main field and causing a shift in the neutral plane.

BRUSHES—Sliding contacts, usually carbon, that make electrical connection to the rotating part of a motor or generator.

COMMUTATION—The act of a commutator in converting generator output from an ac voltage to a dc voltage.

COMMUTATOR—A mechanical device that reverses armature connections in motors and generators at the proper instant so that current continues to flow in only one direction. In effect, the commutator changes ac to dc.

COMPENSATING WINDINGS—Windings embedded in slots in pole pieces, connected in series with the armature, whose magnetic field opposes the armature field and cancels armature reaction.

COMPOUND-WOUND MOTORS AND GENERATORS—Machines that have a series field in addition to a shunt field. Such machines have characteristics of both series- and shunt-wound machines.

CAPACITOR-START MOTOR—A type of single-phase, ac induction motor in which a starting

winding and a capacitor are placed in series to start the motor. The values of X_c and R are such that the main-winding and starting-winding currents are nearly 90 degrees apart, and starting torque is produced as in a two-phase motor.

COUNTER EMF—The voltage generated within a coil by a moving magnetic field cutting across the coil itself. This voltage is in opposition (counter) to the moving field that created it. Counter emf is present in every motor, generator, transformer, or other inductance winding, whenever an alternating current flows.

DELTA—A 3-phase connection in which windings are connected end-to-end, forming a closed loop that resembles the Greek letter Delta. A separate phase wire is then connected to each of the three junctions.

AI-1

DRUM-TYPE ARMATURE—An efficient, popular type of armature designed so that the entire length of the winding is cutting the field at all times. Most wound armatures are of this type.

EDDY CURRENTS—Currents induced in the body of a conducting mass by a variation in magnetic flux.

FIELD—The electromagnet that furnishes the magnetic field that interacts with the armature in motors and generators.

FIELD EXCITATION—The creation of a steady magnetic field within the field windings by applying a dc voltage either from the generator itself or from an external source.

GENERATOR—A machine that converts mechanical energy to electrical energy by applying the

principal of magnetic induction. A machine that produces ac or dc voltage, depending on the original design.

GRAMME-RING ARMATURE—An inefficient type of armature winding in which many of the turns are shielded from the field by its own iron ring.

INDUCTION MOTOR—A simple, rugged, ac motor with desirable characteristics. The rotor is

energized by transformer action (induction) from the stator. More induction motors are used than any other type.

INTERPOLES—Small auxiliary poles placed between main field poles, whose magnetic field opposes the armature field and cancels armature reaction. Interpoles accomplish the same thing as compensating windings.

LAP WINDING—An armature winding in which opposite ends of each coil are connected to adjoining segments of the commutator so that the windings overlap.

LEFT-HAND RULE FOR GENERATORS—A representation of the relationships between motion, magnetic force, and resultant current in the generation of a voltage. The thumb, forefinger, and middle finger of the left hand are extended at right angles to each other. The thumb should point in the direction the conductor moves. The forefinger should point in the direction of magnetic flux from north to south. The middle finger will then point in the direction the generated voltage forces current to flow. Any of three quantities may be found if the other two are known.

MAGNETIC INDUCTION—The generation of a voltage in a circuit by causing relative motion

between a magnetic field and the circuit. The relative motion can be the result of physical movement or the rise and fall of a magnetic field created by a changing current.

MOTOR—A machine that converts electrical energy to mechanical energy. It is activated by ac or dc voltage, depending on the design.

MOTOR LOAD—Any device driven by a motor. Typical loads are drills, saws, water pumps, rotating antennas, generators, etc. The speed and power capabilities of a motor must be matched to the speed and power requirements of the motor load.

MOTOR REACTION—The force created by generator armature current that tends to oppose normal rotation of the armature.

MOTOR STARTERS—Large resistive devices placed in series with dc motor armatures to prevent the armature from drawing excessive current until armature speed develops counter emf. The resistance is gradually removed from the circuit either automatically or manually as motor speed increases.

MULTIPHASE—See polyphase.

POLE PIECES—The shaped magnetic material upon which the stator windings of motors and generators are mounted or wound.

POLE—The sections of a field magnet where the flux lines are concentrated; also where they enter and leave the magnet.

POLYPHASE—Term that describes systems or units of a system that are activated by or which generate separate out-of-phase voltages. Typical polyphase systems are 2-phase and 3-phase whose voltages are 90- and 120-degrees out of phase, respectively. This term means the same as MULTIPHASE.

PRIME MOVER—The source of the turning force applied to the rotor of a generator. This may be an electric motor, a gasoline engine, steam turbine, etc.

ROTATING FIELD—The magnetic field in a multiphase ac motor that is the result of field windings being energized by out-of-phase voltages. In effect, the magnetic field is made to rotate electrically rather than mechanically.

ROTOR—The revolving part of a rotating electrical machine. The rotor may be either the field or the armature, depending on the design of the machine.

SELF-EXCITED GENERATORS—Dc generators in which the generator output is fed to the field to produce field excitation.

SERIES-WOUND MOTORS AND GENERATORS—Machines in which the armature and field windings are connected in series with each other.

SHUNT-WOUND MOTORS AND GENERATORS—Machines in which the armature and field windings are connected in parallel (shunt) with each other.

SLIP—The difference between rotor speed and synchronous speed in an ac induction motor. The rotor will always be slower than the synchronous speed by the amount of slip, otherwise, no voltage would be induced in the rotor.

SLIP RINGS—Contacts that are mounted on the shaft of a motor or generator to which the rotor windings are connected, and against which the brushes ride.

SQUIRREL-CAGE WINDINGS—A type of rotor winding in which heavy conductors are imbedded in the rotor body. The conductors are shorted together at the ends by continuous rings. No insulation is required between the windings and the core. This type of winding is rugged, easily manufactured, and practically maintenance free. It is widely applied in ac induction motors. Physically, it appears as a rotating squirrel-cage, thus the name.

STATOR—The stationary part of a rotating electrical machine. The stator may be either the field or the armature, depending on the design of the machine.

SYNCHRONOUS MOTOR—An ac motor whose rotor is activated by dc. It is characterized by constant speed and requires squirrel-cage windings or some other method to be self-starting.

SYNCHRONOUS SPEED—The speed at which the rotating field in an ac motor revolves. This speed is a function of the number of poles in the field and the frequency of the applied voltage.

AI-3

VOLTAGE REGULATION—A measure of the ability of a generator to maintain a constant output

voltage from no-load to full-load operation. Expressed as a percentage of full-load voltage, the better the regulation, the lower the percent.

WAVE WINDING—An armature winding in which the two ends of each coil are connected to

commutator segments separated by the distance between poles. The winding goes successively under each main pole before reaching the starting point again.

WYE (Y)—A 3-phase connection in which one end of each phase winding is connected to a common point. Each free end is connected to a separate phase wire. The diagram of this connection often resembles the letter Y.

AI-4

MODULE 5 INDEX

Ac generators, 3-1 to 3-4

alternator characteristics and limitations, 3-7

alternator rotors, 3-6, 3-7 basic ac generators, 3-2 frequency, 3-13, 3-14 functions of alternator components, 3-4 to 3-6

parallel operation of alternators, 3-16 practical alternators, 3-4 prime movers, 3-6

principles of ac voltage control, 3-15

rotating-armature alternators, 3-2, 3-3

rotating-field alternators, 3-3, 3-4

single-phase alternators, 3-7, 3-8

summary, 3-16 to 3-21

three-phase alternators, 3-10 to 3-13

two-phase alternators, 3-8 to 3-10
voltage regulation, 3-15, 3-16 Ac motors, 4-1, 4-2
capacitor-start, 4-11
induction, 4-8 to 4-10
resistance-start, 4-12, 4-13
rotating magnetic fields, 4-3
rotor behavior, 4-6, 4-7
series ac, 4-2, 4-3
shaded-pole induction, 4-13, 4-14
single-phase induction, 4-10, 4-11
split-phase induction, 4-11 to 4-13
summary, 4-15 to 4-17
synchronous, 4-7, 4-8
three-phase rotating field, 4-5, 4-6
two-phase rotating field, 4-3 to 4-5 Alternator characteristics and limitations, 3-7 Alternator components, 3-4 to 3-6 Alternator rotors, 3-6, 3-7 Amplidynes, 1-23 to 1-26 Armature losses, dc generators, 1-11
copper losses, 1-12
eddy current losses, 1-12, 1-13
hysteresis losses, 1-13 Armature reaction, 1-8, 1-9
Armature reaction—Continued
compensating windings and interpoles, 1-9, 1-10 Armatures, types of, 2-7, 2-8
drum-wound, 2-8
Gramme-ring, 2-7, 2-8
Commutation, dc generators, 1-7, 1-8 Compound motor, 2-6, 2-7 Compound-wound generator, 1-17, 1-19 Copper losses, armature losses, 1-12, 1-13 Counter emf, dc motors, 2-3, 2-4

D

Dc generators, 1-4 to 1-5
amplidynes, 1-23 to 1-26
armature losses, 1-11, 1-12
armature reaction, 1-8, 1-9
classification of generators, 1-16 to 1-19
commutation, 1-7, 1-8
compensating windings and interpoles, 1-9, 1-10
drum-type armature, 1-14, 1-15
effects of adding additional coils and poles; 1-6
electromagnetic poles, 1-7
elementary dc generator, 1-4, 1-5
elementary generator, 1-2 to 1-4
field excitation, 1-15, 1-16
generator construction, 1-19 to 1-21
Gramme-ring armature, 1-13, 1-14
motor reaction in a generator, 1-10, 1-11
parallel operation of generators, 1-23
practical dc generator, 1-13

safety precautions, 1-26
summary, 1-26 to 1-33
voltage control, 1-21, 1-22
voltage regulation, 1-20, 1-21 Dc motors, 2-1 to 2-3
armature reaction, 2-10, 2-11
compensating winding and interpoles, 2-11
compound, 2-6, 2-7

INDEX-1

Dc motors—Continued
counter ernf, 2-3, 2-4
direction of rotation, 2-9
drum-wound armature, 2-8
Gramme-ring armature, 2-7, 2-8
manual and automatic starters, 2-11
motor loads, 2-4
motor speed, 2-9
practical dc, 2-4 to 2-7
principles of operation, 2-1 to 2-3
series dc, 2-4, 2-5
shunt, 2-5, 2-6
summary, 2-11 to 2-14
types of armatures, 2-7, 2-8 Drum-type armatures, 2-7 Drum-wound armature, 2-8

E

Eddy current losses, 1-12, 1-13 Electromagnetic poles, 1-7

F

Field excitation, 1-15, 1-16 Frequency, ac generators, 3-13, 3-14

G

Generator construction, 1-19 to 1-21 Generators, classification of, 1-16 to 1-19
compound-wound, 1-17, 1-18
series-wound, 1-16
shunt-wound, 1-17 Glossary, AI-1 to AI-4 Gramme-ring armature, 1-13, 1-14, 2-7, 2-8

H

Hysteresis losses, 1-13 I

Induction motors, 4-8 to 4-15 shaded-pole, 4-13, 4-14 single-phase, 4-10, 4-11 speed of single-phase, 4-14, 4-15 split-phase, 4-11 to 4-13

L

Learning objectives, 1-1,2-1,3-1,4-1 ac generators, 3-1 to 3-4 ac motors, 4-1, 4-2 dc generators, 1-4, 1-5 dc motors, 2-1 to 2-3

M

Motor loads, dc motors, 2-4
Motor reaction in a generator, 1-10, 1-11
Motor speed, dc motors, 2-9

P

Parallel operation of generators, 1-23 Prime movers, 1-11, 3-6

R

Rotating-armature alternators, 3-2, 3-3 Rotating-field alternators, 3-3, 3-4 Rotating magnetic fields, 4-3

    rotor behavior in a rotating field, 4-6, 4-7

    three-phase, 4-5, 4-6

    two-phase, 4-3 to 4-5 Rotation, direction of, dc motors, 2-9

S

    Series ac motor, 4-2, 4-3

    Series dc motor, 2-4, 2-5

    Series-wound generator, 1-16

    Shunt-wound generators, 1-17

    Single-phase alternators, 3-7, 3-8

    Single-phase, induction motors, 4-10, 4-11 shaded-pole, 4-13, 4-14 speed of single-phase, 4-14, 4-15 split-phase, 4-11 to 4-13

    Shunt motor, 2-5, 2-6

    Starters, manual and automatic, dc motors, 2-11 Synchronous motors, 4-7, 4-8

T

    Three-phase alternator, 3-10 to 3-13

    three-phase connections, 3-11 to 3-13 Three-phase rotating field, 4-5, 4-6

INDEX-2

Two-phase alternators, 3-8 to 3-10

    generation of two-phase power, 3-8 Two-phase rotating field, 4-3 to 4-5

V

    Voltage control, 1-21, 1-22

    automatic voltage control, 1-23 manual voltage control, 1-21, 1-22 Voltage control, principles of ac, 3-15 Voltage regulation, 1-21, 1-22, 3-15, 3-16

INDEX-3

Assignment Questions

Information : The text pages that you are to study are provided at the beginning of the assignment questions.

ASSIGNMENT 1

Textbook assignment: "Introduction to Generators and Motors," pages 1-1 through 4-18.

1-1. In generators, what principle is used to convert mechanical motion to electrical energy?

    1. Atomic reaction
    2. Electrical attraction
    3. Magnetic repulsion
    4. Magnetic induction

1-2. When you use the left-hand rule for generators, what is indicated by the middle finger?

    1. Direction of flux
    2. Direction of motion
    3. Direction of current flow
    4. Direction of the magnetic field

1-3. The output voltage of an elementary generator is coupled from the armature to the brushes by what devices?

1. Slip rings
2. Interpoles
3. Terminals
4. Pigtails

1-4. An elementary generator consists of a single coil rotating in a magnetic field. Why is NO voltage induced in the coil as it passes through the neutral plane?

1. Flux lines are too dense
2. Flux lines are not being cut
3. Flux lines are not present
4. Flux lines are being cut in the wrong direction

1-5. What components cause(s) a generator to produce a dc voltage instead of an ac voltage at its output?

1. The brushes
2. The armature
3. The slip rings
4. The commutator

1-6. When two adjacent segments of the commutator on a single-loop dc generator come in contact with the brush at the same time, which of the following conditions will occur?

1. The output voltage will be zero
2. The output voltage will be maximum negative
3. The output voltage will be maximum positive

1-7. In an elementary, single-coil, dc generator with one pair of poles, what is the maximum number of pulsations produced in one revolution?

1. One
2. Two
3. Three
4. Four

1-8. If an elementary dc generator has a two-coil armature and four field poles, what is the total number of segments required in the commutator?

1. 8
2. 2
3. 16
4. 4

1-9. How can you vary the strength of the magnetic field in a dc generator?

1. By varying the armature current
2. By varying the speed of armature rotation
3. By varying the voltage applied to the electromagnetic field coils
4. By varying the polarity of the field poles

1-10. Under which of the following conditions does sparking occur between the brushes and the commutator?

1. When operating under normal conditions
2. When there is improper commutation
3. When there is an excessive load current
4. When commutation is in the neutral plane

1-11. Distortion of the main field by interaction with the armature field defines what term?
1. Commutation
2. Mutual reaction
3. Armature reaction
4. Mutual induction

1-12. Distortion of the main field by interaction with the armature field can be compensated for by the use of
1. slip rings
2. interpoles
3. a commutator
4. special brushes

1-13. Motor reaction in a dc generator is a physical force caused by the magnetic interaction between the armature and the field. What effect, if any, does this force have on the operation of the generator?
1. It tends to oppose the rotation of the armature
2. It tends to aid the rotation of the armature
3. It causes the generator to vibrate
4. None

1-14. In dc generators, copper losses are caused by which of the following factors?
1. Reluctance in the field poles
2. Resistance in the armature winding
3. Reactance in the armature and field windings
4. All of the above

1-15. Eddy currents in armature cores are kept low by which of the following actions?
1. Using powdered iron as a core material
2. Limiting armature current
3. Insulating the core
4. Laminating the iron in the core

1-16. What makes the drum-type armature more efficient than the Gramme-ring armature?
1. The drum-type armature has more windings than the Gramme-ring armature
2. The drum-type armature can be rotated faster than the Gramme-ring armature
3. The drum-type armature coils are fully exposed to the magnetic field, while the Gramme-ring armature coils are only partially exposed to the magnetic field
4. The drum-type armature has a laminated core, while the Gramme-ring armature has a solid core

1-17. What type of dc generator application best utilizes the features of the lap-wound armature?
1. High-voltage
2. High-current
3. High-speed
4. Variable-speed

1-18. Which of the following is NOT a major classification of dc generators?

1. Compound-wound
2. Series-wound
3. Shunt-wound
4. Lap-wound

1-19. What characteristic of series-wound generators makes them unsuitable for most applications?
1. They require external field excitation
2. The output voltage varies as the speed varies
3. They are not capable of supplying heavy loads
4. The output voltage varies as the load current varies

1-20. As the load current of a dc generator varies from no-load to full-load, the variation in output voltage is expressed as a percent of the full-load voltage. What term applies to this expression?
1. Gain
2. Voltage control
3. Voltage regulation
4. Load limit

1-21. When two or more generators are used to supply a common load, what term is applied to this method of operation?

1-22. What special-purpose dc generator is used as a high-gain power amplifier?
1. Lap-wound
2. Shunt-wound
3. Amplidyne
4. Compound-connected

1-23. The gain of an amplifying device can be determined by which of the following formulas?
1. GAIN = INPUT + OUTPUT
2. GAIN = INPUT x OUTPUT
3. GAIN = OUTPUT - INPUT
4. GAIN = OUTPUT + INPUT

1-24. The maximum gain possible from an amplidyne is approximately
1. 100
2. 5,000
3. 10,000
4. 50,000

1-25. What determines the direction of rotation of a dc motor?
1. The type of armature
2. The method of excitation
3. The number of armature coils
4. The polarity of armature current and direction of magnetic flux

1-26. When you use the right-hand rule for motors, what quantity is indicated by the extended forefinger?
1. Direction of flux north to south
2. Direction of flux south to north
3. Direction of current
4. Direction of motion

1. Series
2. Compound
3. Split-load
4. Parallel

1-27. Which, if any, of the following situations is a major electrical difference between a dc motor and a dc generator?
1. The armatures are different
2. The shunt connections are different
3. The dc generator requires a commutator, the dc motor does not
4. None of the above

1-28. In a dc motor, what causes counter emf?
1. Improper commutation
2. Armature reaction
3. Generator action
4. Excessive speed

1-29. In a dc motor, how, if at all, does counter emf affect speed?
1. It causes the speed to increase
2. It causes the speed to decrease
3. It causes rapid fluctuations of the speed
4. It does not affect speed

1-30. What is the load on a dc motor?
1. The field current
2. The armature current
3. The mechanical device the motor moves
4. The total current drawn from the source

1-31. When a series dc motor is operated without a load, which of the following conditions occurs?
1. The armature draws excessive current
2. The voltage requirement increases
3. The armature will not turn
4. The armature speeds out of control

1-32. A dc series motor is best suited for which of the following applications?
1. Steady load, low torque
2. Variable load, low torque
3. Steady load, high torque
4. Variable load, high torque

1-33. What is the main advantage of a shunt motor over a series motor?
1. A shunt motor develops higher torque at lower speeds than a series motor
2. A shunt motor can be operated at higher speeds than a series motor
3. A shunt motor draws less current from the source than a series motor
4. A shunt motor maintains a more constant speed under varying load conditions than a series motor

1-34. How can the direction of rotation be changed in a dc motor?
1. Only by reversing the field connections
2. Only by reversing the armature connections
3. By reversing both the armature connections and the field connections
4. 4.By reversing either the armature connections or the field connections

1-35. When the voltage applied to the armature of a dc shunt motor is decreased, what happens to the motor speed?
   1. It becomes uncontrollable
   2. It decreases
   3. It increases
   4. The motor stops

1-36. In a dc motor, the neutral plane shifts in what direction as the result of armature reaction?
   1. Clockwise
   2. Counterclockwise
   3. In the direction of rotation
   4. Opposite the direction of rotation

1-37. The current in the interpoles of a dc motor is the same as the
   1. armature current
   2. field current
   3. total load current
   4. eddy current

1-38. In a dc motor, what is the purpose of the resistor placed in series with the armature?
   1. To counteract armature reaction
   2. To limit armature current
   3. To increase field strength
   4. To prevent overspeeding

1-39. Magnetic induction in an alternator is a result of relative motion between what two elements?
   1. The rotor and the armature
   2. The armature and the field
   3. The field and the stator
   4. The rotor and the field

1-40. Voltage is induced in what part of an alternator?
   1. The commutator
   2. The brushes
   3. The armature
   4. The field

1-41. What are the two basic types of alternators?
   1. Multiphase and polyphase
   2. Alternating current and direct current
   3. Rotating field and rotating armature
   4. Series-wound and shunt-wound

1-42. Which of the following alternator types is most widely used?
   1. Shunt-wound
   2. Rotating-armature
   3. Series-wound
   4. Rotating-field

1-43. The purpose of the exciter in an alternator is to
   1. provide dc field excitation

2. compensate for armature losses

3. compensate for counter emf

4. counteract armature reaction

1-44. An alternator using a gas turbine as a prime mover should have what type of rotor?

1. Turbine-driven

2. Salient-pole

3. Armature

4. Geared

1-45. In alternators with low-speed prime movers, only what type of rotor may be used?

1. Geared

2. Armature

3. Salient-pole

4. Turbine-driven

1-46. Alternators are rated using which of the following terms?

1. Volts

2. Watts

3. Amperes

4. Volt-amperes

1-47. What does the term single-phase mean relative to single-phase alternators?

1. All output voltages are in phase with each other

2. The voltage and current are in phase

3. The phase angle is constant

4. Only one voltage is produced

1-48. In a single-phase alternator with multiple armature windings, how must the windings be connected?

1. Series

2. Parallel

3. Wye

4. Delta

Figure 1A.—Two-phase alternator.

IN ANSWERING QUESTION 1-49, REFER TO FIGURE 1A.

1-49. What is the phase relationship between voltages A and B?

1. In phase

2. 45° out of phase

3. 90° out of phase

4. 180° out of phase

1-50. A two-phase, three-wire alternator has what maximum number of output voltages available?

1. One

2. Two

3. Three

4. Four

THIS SPACE LEFT BLANK INTENTIONALLY.

Figure IB.—Connections for two-phase, three-wire alternator output.

IN ANSWERING QUESTION 1-51, REFER TO FIGURE IB.

1-51. What is the relative amplitude of the voltage at output C as compared to A and B?
1. C is .707 times A or B
2. C is equal to the difference between andB
3. C is 1.414 times A or B
4. C is twice the sum of A and B

1-52. What determines the phase relationship between the individual output voltages in a multiphase alternator?
1. The speed of rotation
2. The number of field poles
3. The method of connecting the terminals
4. The placement of the armature coils

1-53. What is the phase relationship between the output voltages of a three-phase alternator?
1. In phase
2. 60° out of phase
3. 90° out of phase
4. 120° out of phase

1-54. The ac power aboard ship is usually distributed as what voltage?
1. 115-volt, three-phase
2. 115-volt, single-phase
3. 230-volt, single-phase
4. 450-volt, three-phase

1-55. The output frequency of an alternator is determined by what two factors?
1. The number of poles and the number of phases
2. The number of poles and the speed of rotation
3. The speed of rotation and the volt-ampere rating
4. The number of phases and the volt-ampere rating

1-56. A four-pole, single-phase alternator rotating at 18M rpm will produce what output frequency?
1. 60 Hz
2. 400 Hz
3. 1800 Hz
4. 3600 Hz

1-57. Which of the following is the correct formula for determining the percent of regulation of an alternator?
1. Em-En, x 100 = %
Efl
2. Em. x Efl _ <y
100
3- Ehl-Efl x 100 = %
4 - ^ E %

1-58. In most alternators, the output voltage is controlled by adjusting the
1. rotor speed
2. field voltage

3. armature resistance
4. electric load

1-59. When alternators are to be operated in parallel, which of the following alternator characteristics must be considered?
1. Voltage
2. Frequency
3. Phase relationship
4. All the above

1-60. Which of the following motors is/are types of ac motor?
1. Series
2. Synchronous
3. Induction
4. All of the above

1-61. Which of the following types of motors is widely used to power small appliances?
1. Universal
2. Synchronous
3. Polyphase
4. Compound

1-62. A universal motor is a special type of
1. synchronous motor
2. series motor
3. parallel motor
4. polyphase motor

1-63. The number of pole pairs required to establish a rotating magnetic field in a multiphase motor stator is determined by which of the following factors?
1. The magnitude of the voltage
2. The magnitude of the current
3. The number of phases
4. The size of the motor

1-64. In a two-phase motor stator, what is the angular displacement between the field poles?
1. 0°
2. 90°
3. 180°
4. 360°

1-65. Adjacent phase windings of a 3-phase motor stator are what total number of degrees apart?
1. 30°
2. 90°
3. 120°
4. 180°

1-66. Which of the following types of motors has a constant speed from no load to full load?
1. Series
2. Synchronous

3. Induction
4. Universal

1-67. What type of ac motor is the simplest and least expensive to manufacture?
1. Induction
2. Series
3. Synchronous
4. Two-phase

1-68. What term applies to the difference between the speed of the rotating stator field and the rotor speed?

1-69. The speed of the rotor of an induction motor depends upon which of the following factors?
1. The method of connecting the load
2. The dc voltage applied to the rotor
3. The torque requirements of the load
4. The current in the rotor

1-70. What type of ac motor is most widely used?
1. Series
2. Universal
3. Synchronous
4. Single-phase induction

1-71. What type of ac motor uses a combination of inductance and capacitance to apply out-of-phase currents to the start windings?
1. Three-phase
2. Series
3. Synchronous
4. Split-phase induction

1-72. Why are shaded-pole motors built only in small sizes?
1. They have weak starting torque
2. They are expensive in large sizes
3. They are unidirectional
4. They require large starting current

1. Slip
2. Synchronous
3. Rotor error
4. Torque

www.ingramcontent.com/pod-product-compliance
Lightning Source LLC
Chambersburg PA
CBHW081206180526
45170CB00006B/2236